Serotonin Blues

Serotonin Blues

◆

The Roots of Depression

Richard Pray Bonine

iUniverse, Inc.

New York Lincoln Shanghai

Serotonin Blues
The Roots of Depression

iUniverse, Inc.

For information address:
iUniverse, Inc.
2021 Pine Lake Road, Suite 100
Lincoln, NE 68512
www.iuniverse.com

ISBN: 0-595-28204-0

Printed in the United States of America

Contents

Why This Book

Was Stalin out of his mind? And if he was what would it mean to say that the keeper of the keys was mad, the inmates sane? What are we to make of the monstrous crimes of Stalin and other leaders of like kind—Hitler and Mao and—only yesterday—Saddam Hussein?

This question came to mind often as I taught a seminar on the origins and development of modern ideologies at Dalhousie University in the late 1960s. I had taken my bride with me to Halifax on the eastern shore of Canada in the fall of 1964. It was not long before she came to dislike that country intensely because—as she often explained to me—because of the limited selection of shoes. Well shod myself but under some considerable pressure I foolishly agreed to give up my position as associate professor, agreed to leave the barren shoeless wilderness of Nova Scotia behind to return to the shopping paradise just south of the border. I never wrote the scholarly book on ideologies I had thought would be my life's work.

Now, in retirement, I'm writing a long rambling intellectual autobiography that will cover most of the topics from my old seminar. Without footnotes. I may never finish that book but a few sections are already in place, chief among them the essay on mental disorders you are about to read.

I had to write this essay first, before my final analysis of ideologies, because the psychology of rational self-interest still favored by economists and some historians simply won't do. The Holocaust, the liquidation of the kulaks as a class, the Great Leap Forward, the Cultural Revolution, Saddam's torture chambers were all products of deranged minds. If we're to understand the world that was, the world that still is, we will have to take a close look at the inmates of Bedlam.

The Psychiatrists' Manual

The *Diagnostic and Statistical Manual of Mental Disorders*, fourth edition, is a gift to us all from the well-intentioned if unimaginative members of the American Psychiatric Association. In the remarks that follow on insanity I will try to square my vocabulary with the terminology of DSM-IV. So a quick review—perhaps even a generous, an unbiased review—is in order. No footnotes but now and then a page number in parentheses.

The editors decided to stay with the label "mental disorder" for lack of a better despite the implication in that term of an unambiguous division between mental and physical ailments, a division the editors do not accept (xxi). Nor do I.

There is much explanatory and supportive material in the first few chapters of the book and in several appendices, but the main text consists of descriptions of mental disorders grouped into 16 major diagnostic classes, in some but not all cases on the basis of "shared phenomenological features" (9-10). The exceptions to the rule make sense and there's no need to repeat the explanation here.

The editors chose to go with a "categorical classification" rather than a "dimensional system" for their diagnostic classes. "A categorical approach to classification works best when all members of a diagnostic class are homogeneous, when there are clear boundaries between classes, and when the different classes are mutually exclusive." "A dimensional system…works best in describing phenomena that are distributed continuously and that do not have clear boundaries." Nevertheless, "In DSM-IV, there is no assumption that each category of mental disorder is a completely discrete entity with absolute boundaries dividing it from other mental disorders or from no mental disorder" (xxii).

I infer that the Psychiatric Association was driven to adopt a categorical classification for practical reasons, but their members know, and their manual demonstrates, that mental disorders frequently dovetail, often merge one with another, typically differ from the normal or average only in degree.

Each major section opens with a page or two of general remarks on, for example, Dissociative Disorders, followed by descriptions of specific disorders, the name usually preceded by a code, 300.13 Dissociative Fugue for example. If the code is missing from the heading instructions for generating an appropriate code will turn up later.

Without a code the insurance companies won't pay. I saw that in a review somewhere. Honest to God. But I can't find the citation anywhere. No matter. Why be cynical? I don't begrudge psychiatrists—or even psychologists—a living wage. Most of them are real doctors. The contributors to the manual are all listed in an appendix with a label, sometimes PhD but typically MD.

Again I digress. The name of each disorder is followed by a paragraph or two of Diagnostic Features. These paragraphs are generally well written and present no insuperable barriers to your average amateur psychological sleuth. Technical terms are kept to a minimum and explained in context. No need—well seldom any need to reach for the dictionary. Specific criteria for the diagnosis are indicated within the text with letters and numbers, usually A1 to maybe A7, B, C, maybe D.

The criteria, enclosed in a box with the heading Diagnostic Criteria, are repeated in summary form as the last item for a given disorder. The Diagnostic Features are fuller and generally more useful than the Criteria, but not always. Sometimes a more succinct or precise statement in the Criteria clarifies what was hazy in the Features. Read both.

Except for Mental Disorders Due to a General Medical Condition and Substance-Related Disorders the manual has little to say about etiology. Except for a half dozen casual remarks in passing there is *no* theory of mental illness expressed or implied anywhere in the manual. A Reader from Mars would never suspect that Freud had ever puffed a cigar or strolled the streets of Vienna.

The absence of a theory does not bother me in this context. The purpose of the manual is to provide a name and a code for whatever seems to be troubling a client or a patient. So far as I can tell, observing from outside the professional ranks, the manual serves its purpose well.

Which is not to say it can't be faulted. Paul R. McHugh, professor of psychiatry at Johns Hopkins University School of Medicine, is unhappy with the whole enterprise. In his article, "How Psychiatry Lost Its Way" (*Commentary*, December 1999), he notes that "…general medicine abandoned appearance-based classifications more than a century ago. Instead, the signs and symptoms manifested by a given patient are understood to be produced by one or another underlying pathological process."

"In DSM-led psychiatry, however,…psychiatric conditions are routinely differentiated by appearances alone." In DSM-IV McHugh finds some categories of disorder that are real, some dubious, "...and some that are purely the invention of their proponents."

I'm persuaded. But for my purposes it doesn't really matter. What I want from DSM-IV is data. I want to say something about insanity and the insane. Many of my friends and acquaintances qualify but I need a larger sample. I need the Diagnostic Features of DSM-IV.

Diagnostic Features may be followed by Subtypes and/or Specifiers. It strikes me as odd that the editors, in a generally well organized book, have chosen not to reveal what distinguishes one from the other (8). My guess is that Specifiers are qualifiers that indicate degree of intensity, duration and the like, while Subtypes call attention to a distinctive feature.

Here's a string of Specifiers (339-340) that may be used to describe the current Major Depressive Episode in a Mood Disorder: Mild, Moderate, Severe Without Psychotic Features, In Partial Remission, With Catatonic Features, the last of which looks like a Subtype to me. But its their manual.

Here's a Subtype that leaves no room for doubt, Specific Phobia, Animal Type (405-411).

The section on Recording Procedures provides guidelines for names and codes with particular reference to Subtypes and Specifiers. For a Major Depressive Disorder, for example, the first three digits of the code are 296. The fourth is either a 2 to indicate a single Major Depressive Episode or a 3 to indicate recurrent episodes. A 1 in the fifth position indicates mild severity, a 2 moderate severity, and so on (340).

The section on Associated Features and Disorders is often divided into three parts. The first, Associated Descriptive Features and Mental Disorders, contains much valuable detail that does not quite meet the standards for inclusion in Diagnostic Features.

The same section often contains a list of mental disorders associated with the disorder under discussion. Despite the editors' decision to go with a categorical classification it's obvious the mental disorders overlap one another like so many shingles. At the same time, if I may mix my metaphors, the individual disorders protrude like old piles from a sea of anger and anxiety, depression and despair.

Like a physicist hot on the trail of the shy neutrino, the elusive quark, I have found exactly what I was looking for, which is to say grounds for thinking that most mental disorders—excluding those due to a General Medical Condition and those Substance-Related—derive from similar sources, are caused by similar phenomena, have a similar etiology. I will plead my case just as soon as I finish my balanced, objective and unbiased review of DSM-IV.

The second and third parts of Associated Features—Associated Laboratory Findings, Associated Physical Examination Findings and General Medical Conditions—presumably have value for the clinician. For the layman the names of medical conditions, medications and other drugs often make these sections rough going but they're worth a look, even though more often than not the editors feel compelled to enter a disclaimer, for example (323), "No laboratory findings that are diagnostic of a Major Depressive Episode have been identified. However…"

The editors hope to avoid offending ethnic minorities, immigrants and foreigners, a hope that seems to be the main reason for the occasional paragraph on Specific Culture, Age, and Gender Features. You won't find much here that you don't already know about age and gender but now and then something interesting turns up on culture. We learn, for example, of cultures "…in which there is widespread concern about semen loss (e.g., *dhat* syndrome in India)" (447).

Information offered under the head Prevalence is usually vague, never documented. I can't believe this section is useful to anyone.

Information under the head Course [of a disorder] is presumably valuable to the clinician, but for the layman looking to augment his knowledge of mental disorders there's not much here.

Under Familial Pattern we learn there is nearly always a familial pattern, once in awhile an inherited predisposition. "Studies of Anorexia Nervosa in twins have found concordance rates for monozygotic twins to be significantly higher than those for dizygotic twins" (543). I recognized those two long words from a course I took at JC 50 years ago: one-egged and two-egged.

The section on Differential Diagnosis distinguishes one mental disorder from other, often from *many* other remarkably similar disorders. Although I'm skeptical about some of the categories the explanations are usually easy to follow.

And that's the end of my unbiased review of DSM-IV. All that follows is suspect. My first aim is to show that DSM-IV itself tends to support the view that all mental disorders are fundamentally much alike. I don't suggest that the many obvious differences are superficial, but they look to me like different strategies for coping, however unrealistically, with the same basic problems.

In a Glossary of Technical Terms (763-771) DSM-IV defines *affect* as "A pattern of observable behaviors that is the expression of a subjectively experienced feeling state (emotion)."

Toward the beginning of my autobiography, under the heading *Highs and Lows,* I describe affect as a measure of performance and/or behavior. There's nothing like this in DSM-IV but I see my usage as an extension, not a contradiction, of theirs. I will repeat here in summary form what I wrote earlier because before I leave insanity behind I intend to make use of the concept of affect as a measure in an analysis of mental disorders.

A positive affect measures not only individual performance but also generalized behavior in a social setting in which the affection and approval—or disapproval—of others may be crucial. In like manner a negative affect may indicate either a performance off the mark or a false note, a disturbance of some sort, in the social setting.

A negative affect calls for corrective action, but a change of course may be difficult, perhaps beyond the power of an individual. And so negative affects, again like positive, tend to be self-sustaining. They persist where the conditions that give rise to them are not easily altered.

I assume the mechanism that generates affect to measure performance is at least as old as the species, that it guides the individual as he tries to comprehend and master the surrounding world, the world of things.

In the interests of economy the *same* mechanism, I suggest, was gradually adapted to guide the individual in a social world, the world of other people, to measure the behavior of the individual in a group.

Back to DSM-IV. Its definition of affect adds that, "In contrast to *mood*, which refers to a more pervasive and sustained emotional 'climate,' *affect* refers to more fluctuating changes in emotional 'weather.'"

I like the metaphor but I can't accept that mood and affect differ in kind. I would say that mood is whatever affect predominates over a period of time.

The editors of the manual actually see mood and affect more or less my way, if only because they failed to crosscheck their definitions. "Common examples of affect are sadness, elation, and anger." "Common examples of mood include depression, elation, anger, and anxiety." Never mind. Just remember these four words—anxiety, depression, anger, elation—as we drift aimlessly along.

DSM-IV favors the word *dysphoria* but offers no definition in the Glossary. A dysphoric *mood*, however, is defined as "An unpleasant mood, such as sadness, anxiety, or irritability." An irritable mood means "Easily annoyed and provoked to anger." And so, from now on, in the context of this essay, dysphoria is a bowl that holds a mix of depression, anxiety, and anger.

I'm in the mood for moods, dear Reader, simply because you're near me. Funny but when…Let's start with *anxiety*—"The apprehensive anticipation of future danger or misfortune accompanied by a feeling of dysphoria or somatic symptoms of tension. The focus of anticipated danger may be internal or external."

Depression is not listed in the Glossary but you know what it means. I would add that depression suggests resignation, looks to the past. Those who suffer from depression don't *anticipate* misfortune. It's already happened.

In the context of this essay *anger* subsumes any and all forms of hostility and violence whatever the target—rules, laws, conventions, property, other people, one's self. Suicide is in part an expression of anger, a strange phenomenon we'll examine more closely very soon. In the string of quotations coming up in a moment note how often violence toward others is associated with suicide. And watch your daily paper for headlines something like this:

MAN KILLS WIFE, THREE CHILDREN, SHOOTS SELF

Elation, my fourth mood, is much less common than the other three and is usually found just a step ahead of them.

The topic is insanity, but DSM-IV offers a euthymic mood that I'll hold in reserve, a "Mood in the 'normal' range [their quotation marks], which implies the absence of depressed or elevated mood."

Three of the four basic moods—anxiety, depression, anger—if sometimes absent from the Diagnostic Criteria, more often than not turn up as Associated Features. The survey that follows is hardly systematic but will prepare you to believe, at an appropriate time, that all mental disorders in their origins are much the same.

As I prepared a final, clean text of this essay I thought for a moment I would delete most of the quotations that follow. Six pages, I thought, are at least five pages too many. If any psychiatrists should pass this way they would know what to expect while the rest of you could accept my word bolstered by two or three carefully chosen excerpts.

But no, I decided, because I want you not merely to understand in the abstract but to acquire a feeling for the anger and the depression and the anxiety that permeate the mental disorders, for the long shadow of suicide that falls across most categories.

Stuttering:

> *Stress or anxiety have been shown to exacerbate Stuttering. Impairment of social functioning may result from associated anxiety...(64).*

Attention-Deficit/Hyperactivity Disorder:

Associated features…may include low frustration tolerance, temper outbursts, bossiness, stubbornness…demoralization, dysphoria…(80).

Conduct Disorder:

Physical violence may take the form of rape, assault, or in rare cases, homicide (86).

Suicidal ideation, suicide attempts, and completed suicide occur at a higher than expected rate (87).

Schizophrenia:

Dysphoric mood may take the form of depression, anxiety, or anger (279).

Suicide is an important factor [relevant to life expectancy] *because approximately 10% of individuals with Schizophrenia commit suicide (280).*

Schizophrenia, Paranoid Type:

Associated features include anxiety, anger…(287).

The persecutory themes may predispose the individual to suicidal behavior, and the combination of persecutory and grandiose delusions with anger may predispose the individual to violence (287).

Brief Psychotic Disorder:

There appears to be an increased risk of mortality (with a particularly high risk for suicide), especially among younger individuals (302).

Major Depressive Episode:

Individuals…frequently present with tearfulness, irritability, brooding, obsessive rumination, anxiety, phobias, excessive worry over physical health…(323).

The most serious consequence of a Major Depressive Episode is attempted or completed suicide (323).

Manic Episode:

> *Although elevated mood is considered the prototypical symptom, the predominant mood disturbance may be irritability…Lability of mood (e.g., the alternation between euphoria and irritability) is frequently seen (328).*

> *The person may be hostile and physically threatening to others. Some individuals, especially those with psychotic features, may become physically assaultive or suicidal (330).*

> *Mood may shift rapidly to anger or depression (330).*

Major Depressive Disorder:

> *Up to 15% of individuals with severe Major Depressive Disorder die by suicide.*

Bipolar I Disorder:

> *Completed suicide occurs in 10%-15% of individuals with Bipolar I Disorder. Child abuse, spouse abuse, or other violent behavior may occur during severe Manic Episodes or during those with psychotic features (352).*

Specific Phobia:

> *Anxiety is almost invariably felt immediately on confronting the phobic stimulus…(405).*

Obsessive-Compulsive Disorder:

> *The most common obsessions are…aggressive or horrific impulses (e.g., to hurt one's child or to shout an obscenity in church)…(418).*

> *Compulsions are repetitive behaviors…the goal of which is to prevent or reduce anxiety or distress…(418).*

Generalized Anxiety Disorder:

> *Depressive symptoms are also common (433).*

Somatization Disorder:

> *Prominent anxiety symptoms and depressed mood are very common...There may be impulsive and antisocial behavior, suicide threats and attempts, and marital discord (446).*

Body Dysmorphic Disorder:

> *The distress and dysfunction associated with this disorder, although variable, can lead to repeated hospitalization and to suicidal ideation, suicide attempts, and completed suicide (467).*

Dissociative Amnesia:

> *Some individuals with Dissociative Amnesia report depressive symptoms...(478).*

> *Other problems...include sexual dysfunction...self-mutilation, aggressive impulses, and suicidal impulses and acts (478-479).*

Dissociative Fugue:

> *Depression, dysphoria, grief, shame, guilt, psychological stress, conflict, and suicidal and aggressive impulses may be present (482).*

Dissociative Identity Disorder:

> *Usually there is a primary identity that carries the individual's given name and is passive, dependent, guilty, and depressed. The alternative identities frequently have different names and characteristics that contrast with the primary identity (e.g., are hostile, controlling, and self-destructive) (484).*

> *Self-mutilation and suicidal and aggressive behavior may occur (485).*

Depersonalization Disorder:

> *Other common associated features include anxiety symptoms, depressive symptoms...(488).*

Gender Identity Disorder:

> *Adolescents are particularly at risk for depression and suicidal ideation and suicide attempts. In adults anxiety and depressive symptoms may be present (535).*

Anorexia Nervosa:

> *Of individuals admitted to university hospitals, the long-term mortality from Anorexia Nervosa is over 10%. Death most commonly results from starvation, suicide, or electrolyte imbalance (543).*

Primary Insomnia:

> *Symptoms of anxiety or depression that do not meet criteria for a specific mental disorder may be present (553-554).*

Nightmare Disorder:

> *Depressive and anxiety symptoms that do not meet criteria for a specific diagnosis are common among individuals with Nightmare Disorder (580).*

Pathological Gambling:

> *The individual may gamble as a way of escaping from problems or to relieve a dysphoric mood...(616).*

> *Of individuals in treatment for Pathological Gambling, 20% are reported to have attempted suicide (616).*

Adjustment Disorders:

> *Adjustment Disorders are associated with an increased risk of suicide attempts and suicide (624).*

Schizotypal Personality Disorder:

> *Individuals with Schizotypal Personality Disorder often seek treatment for the associated symptoms of anxiety, depression, or other dysphoric affects rather than for the personality disorder features per se (642).*

Antisocial Personality Disorder:

> *Individuals with Antisocial Personality Disorder are more likely than people in the general population to die prematurely by violent means (e.g., suicide, accidents, and homicides) (647).*

Individuals with this disorder may also experience dysphoria...(647).

Borderline Personality Disorder:

> *Completed suicide occurs in 8%-10% of such individuals, and self-mutilative acts (e.g., cutting or burning) and suicide threats are very common (651).*

> *The basic dysphoric mood of those with Borderline Personality Disorder is often disrupted by periods of anger, panic, or despair...(651).*

I hope you haven't forgotten why I laid down this barrage of quotations. I wanted to show you that three of the four most common moods—anxiety, depression, anger—are part and parcel of mental disorders of almost every kind, while the fourth mood, elation, is intimately related to the other three. And of course I also wanted to show how often the culmination of a mental disorder is death by one's own hand.

It's difficult to say how many mental disorders are recognized by the Psychiatric Association but in an Alphabetical Listing of DSM-IV Diagnoses and Codes I count 405 codes. Remember these are spread across 16 diagnostic classes, usually on the basis of "shared phenomenological features" (9-10). I would describe the disorders within most of these classes as variations on a theme. In perhaps the most obvious case—the Anxiety Disorders—the problem is anxiety, cause unknown but tied somehow to any number of things, from elevators to snakes.

The overlapping of mental disorders occurs regularly *across* the 16 diagnostic classes. I'll give just two examples from Associated Descriptive Features and Mental Disorders to illustrate the point.

Bipolar I Disorder:

> *Other associated mental disorders include Anorexia Nervosa, Bulimia Nervosa, Attention Deficit/Hyperactivity Disorder, Panic Disorder, Social Phobia, Substance-Related Disorders (352).*

Paranoid Personality Disorder:

> *In some instances, Paranoid Personality Disorder may appear as the premorbid antecedent of Delusional Disorder or Schizophrenia. Individuals with this disorder may develop Major Depressive Disorder and may be at increased risk for Agoraphobia and Obsessive-Compulsive Disorder. Alcohol and other Substance Abuse or Dependence frequently occur. The most common co-occurring Personality Disor-*

ders appear to be Schizotypal, Schizoid, Narcissistic, Avoidant, and Borderline (635-636).

The section on Differential Diagnosis reveals the same pattern of overlapping and/or merging mental disorders as the Associated Descriptive Features and Mental Disorders. I don't want to quote even one of these lengthy sections in full. Suffice it to say that the clinician must distinguish Posttraumatic Stress Disorder from, or consider an alternate diagnosis of, some 10 to 15 other disorders across half a dozen diagnostic classes (427).

The behavior of people diagnosed with a mental disorder is, to use a term favored by the authors of the manual, maladaptive. The results of maladaptive behaviors are often described in the section on Associated Features. You can easily guess the results so I'll give only one example.

Borderline Personality Disorder:

Recurrent job losses, interrupted education, and broken marriages are common (652).

What is of particular interest to me is that maladaptive behavior of this kind—impairment—is usually required for the diagnosis. This has to mean that you can't tell the crazies from the normals on the basis of their symptoms alone. I understand that the distinction is practical, not theoretical, from the psychiatrists' point of view. But from where I stand it looks like symptoms that *may* indicate a mental disorder exist on a continuum, that the "categorical classification" adopted by the editors of DSM-IV doesn't hold.

The requirement is usually phrased as follows:

The disturbance must cause significant distress or impairment in social, occupational, or other important areas of functioning (Criterion E) (619).

The formula is often tailored to fit a particular disorder but is seldom absent. In those cases where impairment is *not* required for the diagnosis I suspect the patient is plainly ill or obviously headed for jail or bankruptcy. The disorders that caught my eye include Pathological Gambling, Pyromania, Kleptomania, Bulimia Nervosa, Anorexia Nervosa.

And of course sometimes the patient can pass for sane.

Delusional Disorder:

> *Apart from the impact of the delusion(s) or its ramifications, functioning is not markedly impaired and behavior is not obviously odd or bizarre (301).*

Imagine someone—say your father—who might well be diagnosed with a mental disorder if only he were socially or occupationally impaired. But he holds—perhaps in the family, perhaps in some larger organization—a position of relative power that enables him to function effectively and well. He has no need to accommodate others. Others defer to him.

DSM-IV recognizes this situation only with respect to two individuals or, much less frequently, a family. This unequal relationship, once happily named Folie a Deux, is now called a Shared Psychotic Disorder.

> *Usually the primary case in Shared Psychotic Disorder* [commonly diagnosed with Schizophrenia] *is dominant in the relationship and gradually imposes the delusional system on the more passive and initially healthy person.*
>
> *If the relationship with the primary case is interrupted, the delusional beliefs of the other individual usually diminish or disappear.*

I don't fault DSM-IV for not looking beyond the family but obviously similar phenomena can occur in large organizations and institutions such as the contemporary American university and even in entire societies such as the former Soviet Union.

One of the reasons I had to read DSM-IV was the nagging question I posed at the beginning of this essay—was Stalin insane? Were the people who died in the camps thinking Stalin was insane themselves insane? Perhaps, dear Reader, we'll find a satisfactory answer before the bus blows a tire.

As you've probably noticed I'm inclined to give more weight to moods than to odd or bizarre behavior. And DSM-IV itself offers some support for this view.

> **Compulsions** *are repetitive behaviors (e.g., hand washing, ordering, checking) or mental acts (e.g., praying, counting, repeating words silently) the goal of which is to prevent or reduce anxiety or distress, not to provide pleasure or gratification. (418).*

I would go much beyond DSM-IV to suggest that the moods—depression, anxiety, anger, and also, I think, elation with certain qualifications that I hope will soon make an appearance—the moods, I repeat, the moods reflect a problem, a psychological disturbance, while the behaviors and thought processes associated with a mental disorder represent a defense, a strategy that will somehow set things right, stop the pain, banish the mood.

But the defensive measures, the elaborate strategies nearly always fail. Because those who suffer from a mental disorder *cling* to their disease like a lamprey to a coho. Their *present* troubles seem less burdensome than the disaster they fear will overtake them. They hold tenaciously to one set of symptoms to evade another that threatens to supplant the first. I would hazard the guess they all sense suicide at the end of the road.

My sister Carol Lee died not long ago, in November of 2002. In life she was mad as a hatter. But not mad enough to be locked up. She was almost as old as I and had been doctor shopping all her adult life.

She was an intelligent girl and no doubt found it irritating to be reminded by teachers from kindergarten through twelfth grade what a good student big brother had been. She *did* earn a degree from Michigan on the five-year plan, an achievement that perhaps led her to expect more from the world than the world was ready to offer. Certainly she ventured into the labor market at a time when opportunities for women were considerably more limited than they are now. But she never understood that damned near *everyone* has to serve time in the ranks.

She worked for some years as a secretary, mostly in New York, often as a temp, for one fairly long stretch in an interesting position at one of the major theatrical agencies. But she always felt she should be an "executive" secretary, a close associate of movers and shakers, a righthand man, not a servant. The strategy she adopted to advance her career was to paralyze a hand so she couldn't type, a mistake viewed from the outside, perhaps even from the inside because she soon discovered she had to learn to type *very* fast with the other hand alone to retain a job. And retain a job she did for a few more years before returning home to live with mother and father. She was able to persuade the government she deserved a monthly disabilities check.

From the time of the paralyzed hand if not before Carol Lee's goal in life had been to suffer from a medical condition that would preclude employment in a position beneath her dignity, yet would not reflect adversely upon her mental competence. Hence doctor shopping. And ultimately real medical problems, among them Crohn's disease and a ghastly hernia. Toward which the total absence of exercise and a diet of Pepsi Cola and potato chips—sometimes vodka and potato chips—may have contributed.

Carol Lee—like thousands of others I should think—could not face the simple truth and became a master of evasion. She could have been free if only she had said to herself—not me—yes I've been shopping all my life for doctors who would agree with me that my problems are purely physical. But of course she could not, she *would* not do it.

As for myself I cannot, I *will* not believe that fear of exile to the steno pool can account for the stubbornness, the *passion* with which she brandished her medical rationalizations.

Drawing on this brief case history—the only one I have—I want to make two observations I believe are generally applicable to mental disorders, excluding for the moment those Mental Disorders Due to a General Medical Condition and Substance-Related Disorders. My first observation is simply that the predominant mood and the symptoms in any mental disorder emanate from an underlying psychological problem or conflict. The problem is to the mood and the symptoms as cause to effect. A genetic predisposition, a chemical imbalance or the like may be contributing factors.

My sister fled the workplace where she thought herself inadequately respected, insufficiently admired—unloved in a word—and returned home to mother and father who stood ready to offer an abundance of love and affection. At least some of the time. You have to assume there was much more behind the paralyzed hand than a distaste for typing, but there's no need to dig deeper than the homeward flight to make the point.

My second observation follows from the first. Thought—and in particular the *connections* between thoughts—can somehow be kept from consciousness. I have never been happy with the concept of *the* Unconscious—or with any of those other grand abstractions, the Id in particular and the Superego—but that ideas or thoughts can somehow be repressed I have no doubt.

The editors of DSM-IV don't broach the topic. They judge only by appearances. They touch only the surface. But anyone who reads their manual can sense that they know far more than they're willing to say before the general public. Often, *very* often they seem about to share an insight drawn from their vast collective experience but then they stop, cast their eyes downward, hurriedly shuffle off.

Oddly enough, however, they offer the persistent reader a quick look at dissenting opinion. Appendix B contains various Criteria Sets and Axes Provided for Further Study, most of them proposed mental disorders.

Toward the end of the appendix, however, under the head Proposed Axes for Further Study, we find a Defensive Functioning Scale (751-757). These half dozen pages seem to reflect the psychoanalytic point of view although the dread word itself does not deface the page. Most of you will nevertheless recognize several suspect words that *do* make a brief appearance here, in this cul-de-sac, if seldom in the main text—acting out, denial, displacement, projection, reaction formation, repression, sublimation.

I don't think I'll have occasion to make much use of the concepts defined here so I'll leave DSM-IV behind for now to plunge directly into the psychoanalytic swamp.

Man Against Himself

Several decades ago, to celebrate some now forgotten occasion—my birthday perhaps, maybe Christmas—Janet Berne, my mother-in-law, gave me *The Standard Edition Of The Complete Psychological Works Of Sigmund Freud.* I read every word. What sticks in memory is *The Interpretation Of Dreams.* What fascinates me still is the *process* of dreaming.

But right now I want to talk about suicide and I'm going to crib what I need not from Freud but from Karl Menninger, *Man Against Himself,* which I read some 30 or 40 years ago and which I actually re-read in the last few days, the waning days of October, the year 2000.

I know, of course, that most intellectuals now like to say that psychoanalysis has been discredited. Smeared is more like it, but not all the charges leveled against Freud and friends are false.

Are they guilty of reductionism? You bet. Are their methods scientific? Hardly.

But as reclusive and parochial as they are, even a hundred years after they first took the world by surprise they're more exciting—and a lot more fun—than the dog and pigeon people.

In recent years both psychoanalysts and behaviorists have been left choking in the dust by neurologists and pharmacologists and others doing interesting work on the brain and nervous system. It's a pity the leaders in the race show no curiosity about psychoanalysis when they could be rummaging through Freud looking for things to test.

For example, not, as it happens, from *The Standard Edition* but from *The Girl Of My Dreams,* an unpublished work of my own that I would urge you to buy if only it were for sale. Maybe next year.

Everybody dreams every night during every period of REM for rapid eye movement sleep. Subjects aroused from N for Non REM sleep have little to report. Because they're not dreaming. Now put these two thoughts aside for a moment but within reach.

The imagery in dreams, although rarely an exact copy of sensory input, has to come from memory. Is the dream put together on the fly directly from memory during REM sleep? Freud is vague on this point but he specifically repudiates the idea that dream formation depends upon a pool of ideas formed earlier.

He was wrong. I'm telling you the brain chugs along during NREM sleep producing a string of ideas that share a theme or a set of closely related themes but that lack the characteristics of a dream, a string (a linear sequence) from which dreams are formed *later* during REM sleep. Right or wrong it's a neat hypothesis that a clever neurologist might find a way to test.

So much for idle thoughts, for *day* dreams. What is most exasperating about psychoanalysts is that they might have done this kind of experimental work themselves. But they chose instead to spin from a handful of insights a *Weltanschauung* which they then entered as if it were a castle. They have not looked back across the moat these hundred years.

But they're the only folks willing to talk seriously about suicide so let's take a closer look at Menninger. Page references are to the paperback edition, Harcourt, Brace & World, A Harvest Book, 1938. There does not appear to have been a hardcover edition.

In *Man Against Himself* Menninger describes and analyzes not only suicide in the obvious sense but also three closely related phenomena that he names chronic, focal and organic suicide. In chronic or slow suicide the victim, if that's the right word, inflicts pain and suffering upon himself in measured doses to *prolong* suffering so as to *avoid* death.

> *Asceticism, for example, with its varied and ingenious devices for prolonging existence for the purpose of enduring more deprivation, is the very refinement of slow death (77).*

In like manner focal suicide—purposive accidents, self-mutilation, impotence—aims at the destruction of a part or a function of the body in the hope that other parts, other functions can be saved.

In organic suicide—and here Menninger is more tentative—psychological factors of the kind that sometimes lead to chronic or focal suicide may also participate in the development of a physical problem, say high blood pressure, or, in Menninger's best example, boils.

Menninger draws his data from his own clinical experience, from the published reports of other psychiatrists and psychoanalysts, from newspaper accounts and historical writings. Christianity in its first few centuries provides many of his best examples, not, I think, because of any special antipathy toward religion but because the suffering of ascetics and martyrs was much admired and well documented despite protestations against excess from the official church.

You could not squeeze any statistics out of this heterogeneous lump and there *are* those who would condemn newspaper clippings and the lives of the saints as unscientific. Not me. The overall organization of the book is solid, the links between varieties of suicidal experience convincing, the supporting data absolutely compelling.

The theoretical argument less so. But I'll try to offer a fair summary before I select (and alter, perhaps beyond recognition) those concepts that win my favor, satisfy my needs.

Imagine a tank or a vat. Maybe a basin. Yes a basin is better. Filled with two fluids distinct in kind, one negative, one positive, but always mixed one with the other in varying proportions. The mixture ebbs and flows out from, then back into the original basin, but gradually more and more of the mixture—and here I have to change the metaphor—more and more of the mixture is invested in various objects. Sometimes attached to said objects.

The basin, of course, is the self, the individual human being. The negative and positive fluids—Menninger never says "fluid" but often says "flow"—the negative and positive substances that flow have several names, nearly always paired—Thanatos and Eros hence erotic, destructive and constructive, a death instinct, a life instinct, *self*-destructive and creative, hate and love (self-hate, self-love or narcissism), aggression and submission.

Menninger uses these words as if they belonged to one or another of two sets of synonyms. With the result that his use of the word "hate," for example, lends credibility to the very uncertain concept of a "death instinct." Let it go for now. Menninger still has the floor.

The objects are mostly other people but negative and positive substances or impulses can be invested in anything.

For every human being the first object of any importance is the breast. Or the *bottle*, dear Reader. With a nipple larger than life, filled with milk. The breast or the bottle is given to the infant. *And taken away.* By an adult much larger and stronger than your average baby boy or girl.

During the months and years that the baby is suckled, and of course at the end of that period, when the baby is weaned, there is ample opportunity for positive and negative feelings, love and hate, to be invested in the person with the power to give or deny the breast or bottle.

You can hardly talk about a breast in the context of infancy without mentioning a mouth. The infant literally feeds upon the loved object or an extension thereof. This relationship, this connection is the basis or ground for what psycho-analysts call *identification* or, preferably, *introjection*, a process in which an image, an impression, an abstract of one human being is drawn into the mind, the memory, the very being of another. Any subsequent identification bears the mark of the original, carries with it some trace of the infant's ambivalent feelings toward the nurturing but fickle breast. It is introjection that makes it "…possible to treat one's body as if it included the body of someone else" (30).

Two years *before* the publication of Menninger's book Dr. Cole Porter had already developed a remarkably similar theory.

> *I've got you under my skin* [Dr. Porter wrote],
> *I've got you deep in the heart of me,*
> *So deep in my heart, you're really a part of me,*
> *I've got you under my skin.*

Off to the side but still a part of this Freudian model is the conscience, "…an internal, psychological representation of authority, originally and mainly parental authority but fused in later life with prevalent ethical, religious, and social standards" (46). In the psychoanalytic view conscience is "…derived from a portion of the original aggressive instincts which, instead of being directed outward to take destructive effect upon the environment, are converted into a sort of internal judge or king" (47).

Now brace yourself. Here we go.

Joe Basin originally invested a little Thanatos and Eros in himself but in due course he reinvested a fair portion of these elements in others, Eros predominating. He seemed destined for a happy and productive life until the unfortunate day that a close friend—or his wife, his father, a distant cousin, a respected public figure never met—a close friend whose introjected image he carried with him at all times died. Those quantities of Eros and Thanatos once invested in his friend, now lost, were immediately available for reinvestment. But for a number of reasons, among them strong traces of an infantile oral propensity, Joe was not able to find new objects and the detached quantities of Thanatos and Eros returned to their source, Thanatos now predominating. Joe felt a powerful urge to express his aggression against his friend while a harsh conscience demanded propitiation, a life for a life. In the end Joe propped a deer rifle against his temple, pushed the trigger with his thumb, and with a single bullet aimed at the introjected image of his friend he killed himself.

Menninger repeats the above argument throughout the book with slight modifications. In its most succinct form it runs as follows:

> [suicide] *is a death in which are combined in one person the murderer and the murdered (23).*

The formula Menninger seems to favor involves a wish to kill derived from primary aggressiveness, a wish to *be* killed, that is to submit to the dictates of conscience, to accept punishment, and a wish to die which he carefully distinguishes from a wish to be killed (72). Don't worry about the wish to die. It won't be on the quiz.

> *The whole theory of a death instinct and therefore also "the wish to die" element in suicide is only a hypothesis in contrast to the demonstrated facts of the existence of the other two elements (70).*

Freud and his friends had a way with words even when they were off the mark, but "death instinct" was a loser from the beginning. I have not and will not return to the urtext to see what Sigmund said. There's no need. An instinct implies a pattern of behavior, but you and I even as I write are simply wearing out like an old Ford one tune-up away from the scrap heap. And Mother Nature, beaming at the eager awkward boys and nubile girls, doesn't notice, doesn't care.

A life instinct is more plausible than a death drive but I don't think I need one of those either.

I spent the summer of 1957 in Psychico, a suburb of Athens, at the home of John Nicolopoulos, a fellow student at the School of Slavonic and East European Studies. John, his sister Elvi and I, with others who came and went over the summer, often went swimming at Vouliagmeni. Most of the time we swam face down, ping pong balls holding back the Aegean, poking a sea cucumber from time to time, diving for pot shards. We never found anything bigger than the rim of a jug with a handle on one side.

But I once saw something that looked more promising six to eight feet down. I couldn't move it easily, and, excited at the prospect, I planted both feet firmly on the bottom, bent my knees, tried to extricate whatever it was from the coral. A black and red amphora big as a fire hydrant?

Then I remembered where I was, *knew* I was out of breath. Looking up I could see sunlight glancing off the surface not more than five feet away. I was there in a moment gasping for air.

My life had not really been in danger but for some fraction of a second my mind was focused on the shimmering surface of the sea as if death were imminent. No image of Elvi—as nubile as they come—flashed before my eyes. Visions of moussaka danced not through my head.

I should think we can agree there are far too many disparate elements related to the creation and preservation of life to bundle under one name. A life instinct is not a useful concept.

My aim here is not to criticize Freud and Menninger for the practice of inferring from their experiences the existence of some phenomenon, then giving it a name calculated to excite interest, perhaps win adherents. I do it myself whenever I can. It's the shortest route out of the bog. But whatever you invent should be compatible with whatever is known about underlying neurological and physiological structures at the time. And I swear to you, dear Reader, I scan the Minneapolis *Star Tribune* and half a dozen magazines regularly, on the alert for axons and dendrites and neurotransmitters and serotonin and *anything* that might help explain how the mind works. And I *never* invent anything until I've backed myself into a corner from which there's no other escape. I believe in *economy*. The psychoanalysts are big spenders.

I had not remembered and was surprised to read in Menninger that one of the aims of psychoanalytic treatment is to free "the ego from the dominance of the tyrannical super-ego," to replace "conscience by intelligence" (197). I think that will never happen. I certainly hope it won't.

From the Freudian model I accept that conscience is a repository of authority. I also accept that not all the precepts laid down in conscience are available to consciousness. I would add that the precepts embedded in conscience vary from individual to individual and from culture to culture. I'll have more to say on the significance of this variability as we move along.

I have dismissed the life and death instincts as unfounded, implausible, unnecessary. The other terms in the list of substances-that-flow are valid in many contexts, but not in Menninger, not as synonyms. For present purposes only self-destruction demands further analysis. The evidence for the phenomenon is overwhelming. There's no need to justify the term. Unless you feel that a ring through the tongue is a fashion statement.

But if self-destructive behaviors from recklessness to actual suicide are all around us, without something like a death instinct how are we to explain them? Does self-destructive behavior make any sense from any perspective? Why is it that typically *young* people, boys and girls sixteen, seventeen years old, take their own lives?

I think we have to concede that Freud and Menninger got this one right. We need introjection. Or what is much the same thing we need orality. Perhaps now, in the age of Ms. Lewinsky, this irritating and offensive Freudian observation will be more readily accepted. In any case, whatever the contributors to *The New York Review Of Books* may think, the Man in the Street seems to understand the concept. Unless I've misunderstood the bitter and angry expression, "Eat me!"

I don't know how you would accumulate hard evidence for the theory of introjection but Menninger offers many reasons for thinking it is so. I was especially taken by his observation that many would-be suicides don't really want to die.

> *Every hospital interne has labored in the emergency ward with would-be suicides, who beg him to save their lives (23).*

Menninger explains this change of heart in terms of his third element in suicide, an absence of the wish to die. Well yes, but it looks to me as if the would-be suicide has suddenly realized he can't kill the introjected person without killing himself.

What else do you suppose O. P. of Fort Lee, New Jersey, could have been thinking?

> *In Fort Lee, N. J., O. P. wrote two farewell notes, climbed up on the railing of a bridge, ready to jump 250 feet to death. As he teetered, Policeman C. K. shouted: "Get down or I'll shoot." Down got O. P. (64).*

For my next act I need the assimilative mode of thought not as yet mentioned in these particular pages. The assimilative mode—the exclusive mode in dreams, the dominant mode in mental disorders—draws together ideas on the basis of a single theme or set of closely related themes on the one hand, various superficial associations—rhymes for example, or shapes—on the other.

Joe Basin was miraculously given a second chance. This time he invested a hundred dollars from every paycheck in certificates of deposit at 8.6%. He seemed destined for a happy and productive life until the unfortunate day or month or year that he suffered a major loss, the nature of which was not specified. Because it doesn't matter. Joe did not respond well to the crisis. To know why you would have to know the story of his life. Which has not been published. But you can bet he was less mature than most of us. Perhaps with marked traces of an infantile oral propensity.

Joe cast about for someone or something to blame for the loss. It doesn't matter who or what or whether who or what were in any objective sense responsible. Because Joe's mind was now operating in the assimilative mode. Whomever he blamed he saw against the background of the breast or bottle lost or denied so many years ago, now lost or denied again. Joe was in a rage, angry enough, desperate enough to kill. But his conscience could not condone a murder, not of someone he loved. Or even a surrogate for someone he loved. So Joe, hardly knowing what he was about, vented his rage and satisfied his conscience at one and the same time, killing, against all odds, two birds with one stone.

As you can see I've stayed with the Freudian model but have dropped most of the apparatus. And added a twist or two of my own.

Pleasure In Pain

Pain occurs in the mind and may be triggered by anything from a physical injury to a memory. Pain may also be muted or blocked as I think most of us have always known. More than once I've become aware of a minor injury during vigorous exercise from the appearance of blood on a sock or a shirt. No pain had registered in my mind.

According to the current theory, or perhaps I should say according to the *only* theory that has come my way,

> *...pain and other sensations are conceived as "neuromodules" in the brain—something akin to individual computer programs on a hard drive, or to tracks on a compact disk. When you feel pain, it's your brain running a neuromodule that produces the pain experience, as if someone pressed the "play" button on a CD player. And a great many things can press the button. The way Melzack explains it* [Ronald Melzack, Canadian psychologist], *a pain neuromodule is not a discrete anatomical entity but a network, linking components from virtually every region of the brain. Input is gathered from sensory nerves, memory, mood, and other centers, like members of some committee in charge of whether the music will play. If the signals reach a certain threshold, they trigger the neuromodule. And then what plays is no one-note melody. Pain is a symphony—a complex response that includes not just a distinct sensation but also motor activity, a change in emotion, a focusing of attention, a brand-new memory.*

I found this passage in *The New Yorker* for September 21, 1998, "The Pain Perplex" by Atul Gawande. I'm not sure yet how I'm going to exploit this really neat theory but you will note, dear Reader, that it creates plenty of elbow room for Freud and Menninger.

Pain necessarily accompanies all forms of chronic and focal suicide but there is also, in Menninger's view, an erotic element, an "erotization" that draws together pain and pleasure in a peculiar relationship. I don't doubt a connection, but again we face the problem of substances that flow and fuse, the ambiguity of erotization. As Menninger himself concedes—in a footnote on page 26—"This whole matter is, however, still clouded in considerable obscurity..."

I did not anticipate this particular problem. I'll have to improvise.

We have already established, with the aid of Dr. Gawande, that pain *and other sensations* are complex phenomena generated in the mind. Or the brain if you prefer. I think mind is the right word in this context.

Let me list a few sensations I find pleasurable—the taste of eggs fried over easy on a sunny Sunday morning, a warm brick wall against my back on a cool day in early spring, a hot shower, a cold beer, a warm pillow. Evanescent all of them. A prelude to nothing.

Certain sensations associated with sex—a caress, a lingering kiss—are pleasurable in the same way, but only in passing. Because excitement and tension mount to a climax, a release. I mention this not to arouse you from your torpor but to indicate that sexual satisfaction—certainly at the physical, genital level—is not just another pleasurable sensation.

Long before you and I met, dear Reader, I imagined a system that generates affect—creates a mood—that may serve as a measure of performance with respect to some task or other undertaking, or as an evaluation of behavior in a social setting. Sensations may heighten but never determine a mood, and in fact negative correlations are common. A sore throat or a bad knee will not usually spoil a joyous occasion. And a superb dinner may fail to lift the spirits of someone down in the dumps.

I have not explained exactly how my measuring system works. Because I don't know. In the beginning, I should think, it depended on proprioceptors and the like—on basic feedback—but how such a system could be modified and expanded to evaluate behavior in a social setting goes way beyond my scant knowledge of the central nervous system. However it works the measuring system derives from consciousness which I take to be a monitor or scanner.

And now, I fear, I will have to drag conscience into this jerrybuilt structure. For reasons of economy.

The problem that leaps to mind—that's *my* mind. *You* may see more than one problem here, but we're going to approach them one at a time. The problem is that the measuring system tells you how things are going right now (and how they *have* been going for the last several minutes, maybe longer) while conscience tells you what will happen in the future if you don't watch your step. Let me see if I can skate around this hole in the ice.

I will arbitrarily assert that conscience consists of a number of sets of ideas available to memory, all of them evaluated at some time in the past. Evaluated over and over again with such intensity that it's not necessary, perhaps not even possible to re-evaluate them *now*. *That's* how I'll skate around the hole.

The ideas in these sets have to do with prohibitions and requirements laid down by authority—usually parental authority—with an implied or actual threat of a lessening, possibly a total loss of love and security.

Consider a prohibition—do not tell a lie—and a requirement—wash your hands after you go to the bathroom. Now suppose you are tempted to tell a lie, or tempted to rush out of the bathroom without washing your hands. In either case you must make a choice. Or a choice is made for you, determined. Not quite the same thing but let that slide. (Another time, if I'm feeling really lucky, maybe I'll explain freedom.)

All ideas in conscience, whether prohibitions or requirements, carry a high positive value. Make the right choice and the high positive value from conscience sustains or even creates a positive mood. Make the wrong choice and you suffer. Not ordinarily an actual loss—no one is watching—but an anxiety attack. Mild probably. But sometimes not. And your current mood may change for the worse.

Why would anyone make the wrong choice?

Conscience and the measuring/evaluating system—I'm going to call that system the performance and behavior gauge from now on. Sounds more scientific.

Conscience and the performance/behavior gauge develop over time and are unique for each individual. There are those, for example, who adopt lying as a strategy at an early age. If the strategy works—or seems to work—the behavior gauge will register positive when the strategy is employed. Conscience—depending largely, I think, on the context in which the strategy is developed—may or may not follow suit. Conscience, in brief, is constructed or shaped step by step, sometimes in collaboration with authority, sometimes in opposition, but authority operates from a position of strength and some prohibitions, say incest, are more difficult to defuse than others.

The more interesting case is a decision to act against conscience to gain some advantage *expecting* to pay a price, say sweaty palms or a bad night's sleep. People who cheat on exams, or perhaps I should say people who cheated on exams when I was a boy, knew they were guilty, knew there was a price to pay.

The temptation to act against conscience does not release a flood of relevant memories. What reaches consciousness is a vague feeling—no words or images—a feeling only that forces a decision. Do this. Don't do that. No action specified. You know from the context what your conscience requires. The affect has been separated from the original content. I can't begin to explain this in neurological terms but I'm speaking from experience. This is what conscience feels like to me.

What I hope to do next is to describe chronic and focal suicide using the vocabulary and concepts of the last few paragraphs in a manner that would not drive Dr. Menninger to distraction. I'm determined to distinguish pleasure in pain from sexual gratification through pain.

I have no way to know the subjective feelings of ascetics, masochists and the like. I infer from the few remarks that Menninger lets fall a full range of sensations, from an indifference to pain—an insensitivity or numbness—coupled with joy or exultation on the one hand, to full awareness of pain on the other, excruciating pain.

I have never observed a masochist. I assume that when Menninger and others refer to sexual gratification they mean erections, ejaculations, orgasms.

Chronic and focal suicide follow the pattern of an actual suicide—a major loss or more likely a series of minor losses and failures, disappointment, mounting frustration and anger ultimately focused on some introjected person against the background of the original loss, a desire for revenge directed, through the intervention of conscience, against both the introjected object and the self. When death does not ensue from the self-inflicted damage the erotic element is often much in evidence, the reason, I suppose, that psychoanalysts see a *fusion* of the life and death instincts.

Let's look at the simplest case, a physical injury inflicted upon the self, the cutting or burning of one's own flesh. I still find it hard to believe people really do such things but here's something I read just the other day in the Minneapolis *Star Tribune* (11/21/00).

> *Dear Abby: This is in response to "Worried Mother, Chandler, Ariz.," whose daughter is self-mutilating. I had this problem in high school.*

> *Dear Abby: I am almost 17 and have been cutting off and on since I was 13…I will have scars on my arms, back and stomach for the rest of my life.*

> *Dear Abby: I have been a cutter for 15 years.*

Start with the pain. We know that pain is experienced in the mind and can be blocked or muted. But what you or I would experience as pain can never be experienced as pleasure by others. Pain is negative. The highest possible rating is zero.

There are no pleasurable sensations associated with an experience of this kind. What might look like pleasure is a positive *mood*, typically I should think a low-keyed elation, a subpar euphoria.

Two sources contribute to a positive rating for the total experience. The first is a positive boost from conscience, the second a measure of satisfaction from sexual excitation. Whether the satisfaction derives more from contact with the introjected object or the self I'll leave to the psychoanalysts.

Why would the behavior gauge give a positive rating to an experience of this kind? Is this a malfunction? I think not. I think it's more like a design flaw. There is no absolute scale of values. The gauge is a product of each individual's life and for some poor devils the highest available rating looks like hell on earth to those more fortunate.

This view of behaviors that approach but fall short of actual suicide makes sense only if you believe, as Menninger does, as I do, that chronic, focal and organic suicide all represent efforts to stave off a worse fate, the loss of love, the loss of any hope for love, the loss of life. And so I'll close this section with a quotation from Menninger that does not square as well as I would like with either his theory or mine, but that I find moving, humbling.

> *The extraordinary fact that a person should enjoy suffering or should prefer pain to pleasure cannot be easily explained. It can be understood only when one sees that the visible sufferings are far less than the invisible sufferings of such a person, or rather the invisible fears of suffering. To put this very simply: for some individuals it is better to be pitied than to be ignored; it is more terrible to be cast into outer darkness, either in the sense of being left alone and unloved or in the sense of being castrated or dead than to suffer any conceivable amount of pain (139).*

Loss

I have already split the world of people, the social world, from the world of things, but I think now I will readmit certain material objects into the world of people *insofar as they are familiar*, part of the social fabric, things that may fade away, things subject to decay, things that can be destroyed, stolen, left behind, forgotten.

Loss is the primary cause of all mental illness.

There are effective conventional formulae to assuage even the most painful losses, but I'm concerned here with those who cannot or will not avail themselves of the usual remedies, with those who stubbornly choose to follow their own instincts wherever they may lead. Even into the jaws of death.

An actual loss produces depression. An anticipated loss precipitates anxiety. And as DSM-IV notes time after time depression and anxiety go together like Bessie Smith and the blues. Because someone who has suffered a loss may have good reason to anticipate another.

An anticipated loss—say the serious illness, the possible death of a loved one—may depend upon circumstances that no human being has the power to alter. But a loss may also be anticipated because threatened, either explicitly, by some person or persons, or implicitly, through some concatenation of events, of things said or not said, things done or left undone.

A *threatened* loss—the third of my variations on the theme of alienation—ordinarily derives from experiences of childhood and adolescence that leave the adult uncertain of his standing. Am I loved? Am I *worthy* of love?

A threatened loss provokes a wider range of responses than an actual or anticipated loss. The personality disorders offer the best clues to strategies that may be employed in the service of almost any disorder. Some who fear rejection may focus on the fear itself and make "frantic efforts to avoid real or imagined abandonment" (650—Borderline Personality Disorder). Others may react to a threatened loss with "pervasive and excessive emotionality and attention-seeking behavior" (655—Histrionic Personality Disorder). Still others may display an "excessive need to be taken care of that leads to submissive and clinging behavior and fears of separation" (665—Dependent Personality Disorder).

31

In each of these three patterns of behavior anxiety is the dominant mood, often punctuated by outbursts of anger reflecting resentment and hostility. In all three cases the behavior suggests a belief or rather a desperate hope that the threatened loss may be averted.

But there are others who have ceased to believe, who see little or no possibility of being loved and so avoid any form or degree of intimacy. Their lives are characterized by a "pattern of social inhibition, feelings of inadequacy, and hypersensitivity to negative evaluation" (662—Avoidant Personality Disorder). I would have thought the predominant mood would be depression but I find no support for that view in DSM-IV. Links to both Social Phobia, Generalized Type, and Dependent Personality Disorder suggest that the predominant mood must be anxiety.

The social world is the province of adults. The children and adolescents allowed within the gates are on probation. Full membership is the prize for conformity. Or the appearance of conformity. Rejection is the punishment for those who *will* not conform or *cannot* acquire the requisite skills, a distinction often difficult to draw in practice. Those who can't read are seen as unwilling to conform. And more often than we care to acknowledge we, the adults, tell the child, "I hear you knocking but you can't come in."

I propose—cheating only a little—a new concept, the concept of virtual loss, the loss of or denial of full rights and privileges in the social world. Those who suffer this loss tend to be less intelligent than average or disadvantaged in some other way. Their predominant mood is anger. Their behavior varies from bad to very bad—say from rude to homicidal—but not, as most of us see it, really crazy. But crazy they are all the same.

Loss. Actual, anticipated, threatened and virtual loss. Let's see how these variations on a theme play against the sixteen major categories of DSM-IV. The editors don't number their categories but I will follow my first reference to each with a number in brackets just so you'll know how far we've come, how far we have to go.

Among the many Disorders Usually First Diagnosed in Infancy, Childhood, or Adolescence {1} we find Mental Retardation which I will pass. I see why psychiatrists need the category but the mentally retarded are not insane. Neither are most of those who suffer from a Learning Disorder but they *are* at risk for insanity. We better take a look.

There may be underlying abnormalities in cognitive processing (e.g., deficits in visual perception, linguistic processes, attention, or memory, or a combination of these) that often precede or are associated with Learning Disorders...Although genetic predisposition, perinatal injury, and various neurological or other general medical conditions may be associated with the development of Learning Disorders, the presence of such conditions does not invariably predict an eventual Learning Disorder, and there are many individuals with Learning Disorders who have no such history. Learning Disorders are, however, frequently found in association with a variety of general medical conditions (e.g., lead poisoning, fetal alcohol syndrome, or fragile X syndrome) (47).

The phrase "no such history" suggests that *some* Learning Disorders may be primarily mental or psychological. Now DSM-IV always assumes, as do I, that there's a physical substratum behind—or rather beneath—all mental and psychological phenomena. I mention this because the DSM-IV editors are more sensitive than I to political currents. As a result of which their prose is sometimes more obscure than it need be. As in the lengthy passage just quoted.

One of the diagnostic requirements for a Learning Disorder also suggests a mental or psychological problem. "If a sensory deficit is present, the learning difficulties must be in excess of those usually associated with the deficit" (46-47).

The first and most important of the Learning Disorders is Reading Disorder or dyslexia. Its essential feature "...is reading achievement...that falls substantially below that expected given the individual's chronological age, measured intelligence, and age-appropriate education (Criterion A)" (48). "Substantially below" is a statistical measure "...usually defined as a discrepancy of more than 2 standard deviations between achievement and IQ" (46). Which doesn't tell us a hell of a lot about the nature of the problem.

Donna Renshaw, wife of Quintus—together proprietors of the Pentwater Inn where I spend the month of August every year—has a dyslexic son now highly successful as a house painter, mostly interiors. I've often heard Donna describe to guests of the inn during breakfast the hours she spent helping her son through school. And in fact most successful men have been dyslexic, overcoming adversity, achieving greatness, only through the patient untiring help of a beloved mother. Or father. Or dedicated, saintly teacher. Just read the paper.

Hawks I've been told can see a mouse move in the grass two hundred feet below. This remarkably keen vision suggests to some of us an evolutionary development over millennia if not eons.

Homo sapiens, by way of contrast, learned to read overnight. The leap from tracks in the sand, the telltale signs of spoor, to indentations in a clay tablet, to hieroglyphics down a scroll of papyrus, did not depend upon an evolutionary modification of the eye.

Reading a newspaper is not a natural act. Put comprehension, understanding aside. The *physical act* of reading, the concentration, the focus, the fine control, is demanding, even in the absence of any physical "deficits" as DSM-IV would phrase it.

For the lucky few learning to read is great fun. For many—for the majority I suspect—a chore. For a small but significant minority a daunting task, a torment.

"The prevalence of Reading Disorder in the United States is estimated at 4% of school-age children" (49). With the support and help of concerned parents and teachers some of these children learn to read tolerably well. Or learn to avoid situations in which reading is required. Some never learn.

In descriptions of the Reading and other Learning Disorders there are no references to anger, but many of those with a learning disability will sooner or later be diagnosed with something else—say Conduct Disorder, or perhaps Oppositional Defiant Disorder—in which anger is the predominant mood.

In brief we're talking about resentful boys and girls who cannot or perhaps will not, who in any case *do* not acquire certain basic skills and *for that reason* are, in effect, rejected, for that reason denied access to the social world, for that reason made to suffer an irreparable virtual loss.

DSM-IV describes a Mathematics Disorder and a Disorder of Written Expression, both similar to Reading Disorder, neither ordinarily found in the absence of Reading Disorder. The problem in all three cases is basically the same.

(DSM-IV always provides a default diagnosis for the odd set of symptoms, in this case the fourth Learning Disorder—Learning Disorder Not Otherwise Specified. I won't mention these again. You'll probably have to subtract 1 from any count I give. Or maybe add 1.)

DSM-IV saves space for a group—Motor Skills Disorder—with only one member—Developmental Coordination Disorder. "The essential feature…is a marked impairment in the development of motor coordination (Criterion A)" (53). The diagnosis may *not* be made if the coordination difficulties are due to a general medical condition such as cerebral palsy. Younger children with this disorder "…may display clumsiness and delays in achieving developmental motor milestones (e.g., walking, crawling, sitting, tying shoe laces, buttoning shirts, zipping pants)."

DSM-IV doesn't offer much data on this topic but I see a similarity to the Learning Disorders. I see difficulties with respect to the acquisition of basic skills, but in this case related to tasks that are *not* inherently difficult. Except maybe for tying shoe laces. I don't remember. I can see an image of children sprawled across the floor putting on galoshes. Is that from *my* childhood?

The diagnosis is made only if this impairment significantly interferes with aca-demic achievement or activities of daily living (Criterion B).

If it's that much trouble to button your shirt I have to assume early signs of resistance to the discipline demanded by the civilized world, resistance that becomes much more obvious in disorders associated with adolescents.

DSM-IV describes five Communication Disorders. The first of these—Expressive Language Disorder—is similar to Reading Disorder. The diagnosis in both cases depends on test scores.

> *Expressive Language Disorder may be either acquired or developmental. In the acquired type, an impairment in expressive language occurs after a period of normal development as a result of a neurological or other general medical condition (e.g., encephalitis, head trauma, irradiation). In the developmental type, there is an impairment in expressive language that is not associated with a neurological insult of known origin (56).*

That last phrase suggests that psychiatrists may believe but cannot be certain there's a neurological basis for the disorder. Which sounds plausible to me. But I would stress the inherent difficulty in mastering a language. Yes, I know, every-one learns to speak, but few learn to speak well. And look at the deficiencies in the speech of those with Expressive Language Disorder.

> *These features* [the linguistic features of the disorder] *include a limited amount of speech, limited range of vocabulary, difficulty acquiring new words, word-finding or vocabulary errors, shortened sentences, simplified grammatical structures, limited varieties of grammatical structures (e.g., verb forms), limited varieties of sentence types (e.g., imperatives, questions), omissions of critical parts of sentences, use of unusual word order, and slow rate of language development (55-56).*

I don't plan to comment on disorders due to a general medical condition on the grounds that those who suffer from them are not crazy, which is obviously true in the case of someone with Alzheimer's disease. But in the context of language difficulties in childhood that rule of thumb is not always applicable. Children who experience great difficulty learning to speak, whatever the reason, are at risk for a major virtual loss, for rejection, for a troubled and unhappy life.

Children diagnosed with Mixed Receptive-Expressive Language Disorder experience difficulty not only in expressing their own thoughts but in understanding the thoughts of others. Just like Expressive Language Disorder. Only worse.

Phonological Disorder, the third of the Communication Disorders, has to do mostly with errors in articulation.

> *Severity ranges from little or no effect on speech intelligibility to completely unintelligible speech...The most frequently misarticulated sounds are those acquired later in the developmental sequence (l, r, s, z, th, ch)...Lisping (i.e., misarticulation of sibilants) is particularly common (61).*

> *Although there may be an association with clear causal factors such as hearing impairment* [and others named]*...at least 2.5% of preschool children present with Phonological Disorders of unknown or suspect origin, which are often referred to as* functional *or* developmental *(62).*

Clear causal factors or not I would stress the inherent difficulties children must overcome to master the spoken language. The observation from DSM-IV that follows looks to me like problems solved and happily left behind.

> *In mild presentations with unknown causes, spontaneous recovery often occurs (62).*

Let me digress for a moment to propose a new disorder for inclusion in DSM-V, the Punctuation Disorder. The essential feature of Punctuation Disorder is the stubborn willful *maddening* insertion of commas into a text where they serve *no useful purpose* (Criterion A).

In Stuttering, the fourth Communication Disorder, the problem arises in part because of the inherent difficulty of the task, but stuttering is also plainly a result of pressure to perform up to standard, pressure to meet the entrance requirements of the social world.

> *The extent of the disturbance varies from situation to situation and often is more severe when there is special pressure to communicate (e.g., giving a report at school, interviewing for a job). Stuttering is often absent during oral reading, singing, or talking to inanimate objects or to pets (63).*

> *Some research suggests that up to 80% of individuals with Stuttering recover, with up to 60% recovering spontaneously. Recovery typically occurs before age 16 years.*

Stuttering looks like a minor problem compared to the other Learning and Communication Disorders, a problem often solved. Perhaps I should state explicitly that I don't believe in "spontaneous recovery." The child must learn to cope with the pressures brought to bear before he can control his speech.

DSM-IV describes five Pervasive Developmental Disorders, the most familiar of them probably Autistic Disorder.

> *These disorders…are often associated with some degree of Mental Retardation…*

> *The Pervasive Developmental Disorders are sometimes observed with a diverse group of other general medical conditions (e.g., chromosomal abnormalities, congenital infections, structural abnormalities of the central nervous system) (65-66).*

Some of the children who suffer from these disorders may look crazy some of the time but I think they are not. There is some physical explanation—probably a genetic defect—for their behavior.

I had thought I could get away with this artless distinction—the crazies on one side, those with a physical disease or disability on the other—but a problem lies in wait for me in the section on schizophrenia which some authorities consider an illness like measles or chickenpox. I won't buy the chickenpox analogy but I don't have to deny a genetic predisposition or other physical cause as contributing factors in mental disorders. In my remarks on schizophrenia I will stress the content of delusions and hallucinations. Because delusions and hallucinations reveal to the neighbors that the character down the street is out of his mind. To the best of my knowledge autistic children do not have delusions.

Five disorders—three that really count—are described under the heading Attention-Deficit and Disruptive Behavior Disorders. The first of these is the wildly popular Attention-Deficit/Hyperactivity Disorder known to the press but not to the Psychiatric Association as ADHD.

This disorder is really two in one. Among the criteria for a diagnosis on the Attention-Deficit side I find the following most interesting.

> *Tasks that require sustained mental effort are experienced as unpleasant and markedly adversive. As a result, these individuals typically avoid or have a strong dislike for activities that demand sustained self-application and mental effort or that require organizational demands or close concentration (e.g., homework or paperwork) (Criterion A1f) (78).*

On the Hyperactivity side children with this disorder can drive both adults and other children right up the wall.

> *Individuals with this disorder typically make comments out of turn, fail to listen to directions, initiate conversations at inappropriate times, interrupt others exces-*

sively, intrude on others, grab objects from others, touch things they are not sup-
posed to touch, and clown around (79).

Additional material in the Associated Descriptive Features and Mental Disorders reveals how serious the disorder can sometimes be.

Academic achievement is often impaired and devalued, typically leading to conflict
with the family and school authorities. Inadequate self-application to tasks that
require sustained effort is often interpreted by others as indicating laziness, a poor
sense of responsibility, and oppositional behavior. Family relationships are often
characterized by resentment and antagonism, especially because variability in the
individual's symptomatic status often leads parents to believe that all the trouble-
some behavior is willful (80-81).

Perhaps because it is. DSM-IV insists that the diagnosis *not* be made if the observed behavior is due to a "primary oppositional attitude" rather than "difficulties with attention" (78). But who knows what "primary oppositional attitude" means? Or how to distinguish it from "secondary oppositionalism" which the diagnostic criteria permit?

Oppositionalism?

The profession is pandering to the *Zeitgeist*. In most cases the observed behavior *does* indicate hostility to the restraints imposed, the demands made by parents and teachers, the representatives of civilization from an adolescent perspective.

I have to believe the members of the Psychiatric Association were once children themselves. Don't they remember the horrors visited upon substitute teachers by the best behaved of boys and girls?

I took trigonometry in the eleventh grade. Our teacher was a student from nearby Calvin College, an intelligent and personable young woman but a *practice* teacher, not a *real* teacher. We never gave her a chance. And to this day I can't tell a cosine from a hat rack.

For a few memorable months in the school year 1960-1961 I served as a study hall supervisor at Ottawa Hills, my old high school. I was not strict about observing the rules when only 30 to 40 students were scattered around the cafeteria, but at certain hours two to three hundred boys and girls swarmed into the now cramped space where I presided alone and unarmed. I once foolishly touched a disruptive boy lightly on the shoulder thinking I might persuade him to settle down. The next moment we were on the floor wrestling. I was happy to find employment elsewhere.

During that same year I spent several days as a substitute at a junior high, Harrison I think, teaching seventh grade arithmetic and ninth grade civics. The ninth graders were sullen and uncommunicative. I tried to follow the lesson plan, tried to engage them in a discussion of the Bill of Rights, but mostly I prayed for the hour to end. The seventh graders were still childish, still fun to be with, but they had their own agenda, mostly chattering. They didn't have the time for fractions.

Children *will* test you. They *will* discover what the limits are. You better be ready and you better be strong.

Children from any background may be diagnosed with ADHD but a high proportion seem to have been born into unfavorable circumstances.

> There may be a history of child abuse or neglect, multiple foster placements, neuro-toxin exposure (e.g., lead poisoning), infections (e.g., encephalitis), drug exposure in utero. low birth weight, and Mental Retardation (81).

In most boys and girls diagnosed with ADHD "symptoms attenuate during late adolescence and adulthood" (82) which doesn't look like a favorable prognosis to me. Certainly some are fated to suffer from a more serious mental disorder later in life.

From my perspective children diagnosed with Attention-Deficit/Hyperactivity Disorder have already suffered or are in imminent danger of suffering a grave virtual loss, a lessening if not a denial of love and security, an interminable probation if not exclusion from the ranks of the grownups.

The words I have chosen—denial, exclusion—place the responsibility for broken lives on parents and teachers but certainly some of the children are intractable. There is something in all of us that resists the yoke of civilization. And the very young cannot know that successful resistance means ultimate failure.

ADHD has two companion disorders, Oppositional Defiant and Conduct. The first has to do mostly with attitude, the second with action, aggressive action directed against property, animals, other people. In other respects these two disorders are much like one another and much like ADHD, with an obvious progression as children grow older from troublesome (ADHD) to irritating and offensive (Oppositional Defiant Disorder) to dangerous (Conduct Disorder).

The more time I spend with DSM-IV the more puzzled I am by the tireless efforts of the editors to distinguish from one another disorders that are very much alike. Disorders that DSM-IV *knows* are very much alike. In the quotations that follow I've used the initial letters instead of the full names of the three "disruptive" disorders to prevent Carpal Tunnel Disorder.

A substantial proportion of children referred to clinics with ADHD also have ODD or CD (81).

ADHD is common in children with CD (81).

When the individual's pattern of behavior meets the criteria for both CD and ODD, the diagnosis of CD takes precedence and ODD is not diagnosed (89).

When criteria are met for both ADHD and CD, both diagnoses should be given (89).

ADHD is common in children with ODD (92).

In a significant proportion of cases, ODD is a developmental antecedent to CD (92).

Because all of the features of ODD are usually present in CD, ODD is not diagnosed if the criteria are met for CD (93).

When the two disorders [ADHD, ODD] co-occur, both diagnoses should be made (93).

All three are often associated with or lead to Learning and Communication Disorders, Substance Related Disorders, Antisocial Personality Disorder, Mood and Anxiety Disorders.

The range of aggressive behavior characteristic of boys—and a few girls—diagnosed with Conduct Disorder is staggering.

They may display bullying, threatening, or intimidating behavior (Criterion A1); initiate frequent physical fights (Criterion A2); use a weapon that can cause serious physical harm (e.g., a bat, brick, broken bottle, knife, or gun (Criterion A3); be physically cruel to people (Criterion A4); or animals (Criterion A5); steal while confronting a victim (e.g., mugging, purse snatching, extortion, or armed robbery (Criterion A6); or force someone into sexual activity (Criterion A7). Physical violence may take the form of rape, assault, or in rare cases, homicide (86).

More interesting and revealing than the actual aggressive behavior, I think, is the attitude toward others. Boys and girls diagnosed with Conduct Disorder are deceitful. They lie. They break promises. They "…may readily inform on their companions and try to blame others for their own misdeeds." "They may be callous and lack appropriate feelings of guilt or remorse." They may dissimulate, *expressing* guilt in an effort to avoid punishment. In sum they "…may have little

empathy and little concern for the feelings, wishes, and well-being of others"
(87).

These are boys who have not entered and may never enter the world of adults,
boys who have failed or been rebuffed time and again, boys who have endured an
endless series of virtual losses. Their predominant mood is anger and they strike
out blindly at targets of convenience as if they could set the world right with a bat
or a broken bottle.

They may enjoy an occasional small triumph. They may believe they can fool
the authorities, beat the system. And in the short run maybe they can. The performance and behavior gauge shaped by these experiences may give a positive
reading for actions that are basically self-destructive.

Conscience presupposes family and community and in their virtual absence
remains rudimentary, offers no warning of the consequences that must follow the
habitual maltreatment of all and sundry.

Some of these boys—a majority—adjust as they grow older. DSM-IV offers
no explanation. I imagine that a job away from home and school and encounters
with employers and others who cannot be bullied or conned usually make the difference.

The others soon pass a point beyond which they are doomed. Their internal
guides, conscience and the behavior gauge, if operative at all, point in the wrong
direction. As they grow older they find the world less tolerant. Small triumphs are
rare. Anger, now colored by depression, grows more intense. Ahead lies only
trouble. At age 18 the diagnosis may change, say from Conduct Disorder to Antisocial Personality Disorder, but the prognosis remains unfavorable in the
extreme.

Some of these boys take their own lives. Their behavior conforms fairly well to
the analysis I've borrowed and adapted from Menninger. A series of losses.
Mounting anger directed at legions of introjected objects assimilated one to
another. The urge to kill redirected by conscience against the self. I'm forced to
assume that even a weakened conscience, a rudimentary conscience can prohibit
murder. But no, I'm not happy with this obvious flaw in the argument. If
Microsoft sends me a fix I'll let you know.

I'm going to race right by Feeding and Eating Disorders of Infancy or Early
Childhood. They may have something to do with orality. But probably not. And
two of the three—Pica and Rumination Disorder—tend to induce a certain
queasiness in your author. Only pediatricians need to know about these dietary
deviations.

But I *did* promise anecdotes. Many decades ago—say 1937—I watched Douglas, the boy across the street, remove an earthworm from a birdbath in his backyard and eat it. I did not join in the repast.

I'd like to skip Tic Disorders too. They don't make much sense. But I better not.

Tics can be motor or vocal, simple or complex. DSM-IV sorts out the possible combinations and finds four Tic Disorders, all basically the same.

A *vulnerability* to a Tic Disorder is inherited (102-103) but the particular manifestation in a given case is plainly mental and can be suppressed for a time.

> It [a tic] *is experienced as irresistible but can be suppressed for varying lengths of time. All forms of tic may be exacerbated by stress and attenuated during absorbing activities (e.g., reading or sewing). Tics are usually markedly diminished during sleep (100).*

Facial twitches were all I knew before reading DSM-IV but Tourette's Disorder is the Baskin-Robbins of tics. The complexity of some—deep knee bends, retracing steps, twirling when walking—the content of others—yelps, barks, the uttering of obscenities—all suggest a mental component.

Coprolalia, the uttering of obscenities, may be involuntary, but no one is born with a vocabulary of four-letter words. At some level the speaker knows what he's saying. And he doesn't like the constraints imposed by a fastidious civilization.

Tics are a form or variety of self-destructive behavior that can produce serious injury. Head banging, for example, may result in blindness due to retinal detachment.

But most tics, I suspect, express indirectly the fear of an anticipated loss. I see a link with the anxiety disorders, especially with obsessions and compulsions commonly associated with tics (102). DSM-IV insists that tics "...are not aimed at neutralizing the anxiety resulting from an obsession" (101). Fair enough. But I didn't say that tics neutralized anxiety. I said they expressed fear of an anticipated loss.

The *sine qua non* of civilization is strict control of defecation and—with some leeway—urination. The Elimination Disorders—Encopresis and Enuresis—suggest a certain reluctance to conform to the standard, a packet of hostility toward those who enforce it, often combined with some problem, say constipation, that makes conformity more difficult.

Spontaneous remission is the norm for Enuresis but Encopresis may persist and may be associated with Oppositional Defiant or Conduct Disorder, that is with more than a certain reluctance to conform.

My son Michael was diagnosed shortly after birth with a heart problem, a ventricular septal defect was the medical term I think, but in any case a hole located so that a significant portion of his blood was pumped directly back into the heart instead of out into the body. Michael was not a blue baby but the prognosis was a gradual enlargement of the hole, an increasing burden on the heart, probably death within ten to twelve years.

In May or early June of 1972, not long before his fifth birthday, my wife Sue and I took Michael to the Mayo Clinic for open heart surgery. The operation was a complete success but who knows what impact it had on Michael's mind. There's no way to tell a boy not yet five that you're going to let someone cut through his breastbone with a buzz saw.

I was in Michael's hospital room one morning not long after the operation. He seemed all right. There were three or four young women from the Philippines in the room—aides or nurses in training—laughing and talking with Michael and the nurse on duty when a doctor on rounds entered with a false cheery greeting that did not sit well with his patient. I can't be sure of the exact words but Michael, sitting up straight in bed, in a loud firm voice said something like this—I hate doctors. I hate nurses. I hate mommy. I hate daddy.

I suppose he still hates daddy. I haven't seen him since the summer of 1988, the summer before his senior year at Oberlin. There've been a few signs of affection. In a secondhand bookstore in Mexico City he found and sent me a copy of *The Dial* for November of 1922 that includes the first published version of *The Waste Land*. For a time we communicated by email, but he's turned against the computer my sister tells me, and once again an abyss has opened between us.

I'm happy to say I can update the preceding paragraph. Michael and I spent most of two or three days together at the time of my sister's funeral. There seemed to be warm feelings all around. He still doesn't answer my letters but I try not to push my luck.

Soon after the incident in the hospital room I left Rochester for Pentwater to open our summer home. For the last few days of Michael's stay at the Clinic Sue was joined by Marybelle Bentley, an old friend from Grinnell College. On the day Michael was discharged from the hospital the three of them had lunch *al fresco* at a nearby restaurant. Sometime during lunch Michael left the table, stepped into a small plot of ground surrounding a tree, dropped his pants, then dropped a heavy load. Never before had he done anything like this, never since.

The message was clear. I imagine it still reverberates in the lonely chambers of his mind.

The very long section on the disorders of infancy and childhood closes with four disorders that apparently stand alone. I really like the first of these—Separation Anxiety Disorder—because it fits so well into the scheme of things I'm proposing. What is Separation Anxiety but the fear of an anticipated loss?

I also like this disorder because the children are naive and in their apprehensions and beliefs you can see the origins of some of the symptoms characteristic of adults. The Diagnostic Features are obvious—anxiety concerning separation—so I'll quote from the Associated Features.

> *Depending on their age, individuals may have fears of animals, monsters, the dark, muggers, burglars, kidnappers, car accidents, plane travel, and other situations that are perceived as presenting danger to the integrity of the family or themselves. Concerns about death and dying are common.*

> *Children may complain that no one loves them or cares about them and that they wish they were dead.*

> *When alone, especially in the evening, young children may report unusual perceptual experiences (e.g., seeing people peering into their room, scary creatures reaching for them, feeling eyes staring at them (111).*

The Course of the disorder is of particular interest to me.

> *Separation Anxiety Disorder may develop after some life stress* [after some **loss**] *(e.g., the death of a relative or pet, a change of schools, a move to a new neighborhood, or immigration (112).*

Selective Mutism is the refusal to speak—DSM-IV says *failure* to speak—in certain social situations by children who can and do speak normally in other situations. This disorder usually fades away in a few months and I infer from the Associated Features that it's a direct and obvious resistance to socialization coupled sometimes with a fear of rejection for poor performance.

> *Associated features of Selective Mutism may include excessive shyness, fear of social embarrassment, social isolation and withdrawal, clinging, compulsive traits, negativism, temper tantrums, or controlling or oppositional behavior, particularly at home (114).*

Reactive Attachment Disorder is associated with "grossly pathological care" (116) much worse than what I've called virtual loss—"persistent disregard of the child's basic physical needs" or "repeated changes of primary caregiver that prevent formation of stable attachments" (116). The child has no chance unless "an appropriately supportive environment is provided" (117).

Stereotypic Movement Disorder is usually associated with Mental Retardation. I'll ignore it to hasten my exit from this dreary review of childhood disorders.

The appearance of the term *organic mental disorder* in DSM-III-R for revised was a source of acute embarrassment not to say anguish for the editors of DSM-IV because it implied that *nonorganic* mental disorders did not have a biological basis (123). And you know how unscientific *that* was.

The solution that suggests itself to me for this embarrassing problem is to announce unambiguously that yes, of course, there *has* to be a biological basis for *all* mental disorders, but as it happens I'm interested primarily in crazy people, in *mental* cases.

And rather surprisingly, after much backing and filling, DSM-IV proposes the *same* solution, reserving the term *primary mental disorder* for those conditions that have no specified etiology (165), i.e., for the crazy people, the *mental* cases.

All those conditions that had been described in DSM-III-R as organic mental disorders are described, in DSM-IV, as either due to a general medical condition or substance-related. Disorders in these *two* classifications appear in one of the manual's *three* major categories. And in most of the others. You don't have to know all this. I'm just telling you I really read the manual.

The *three* major categories are:

Delirium, Dementia, and Amnestic and Other Cognitive Disorders [2]. This category includes disorders due to a general medical condition *and* substance-related *or* a combination of the two. No cross references.

Mental Disorders Due to a General Medical Condition [3]. A very short section with a few cross references, e.g., Anxiety Disorder Due to a General Medical Condition. Text and criteria are included in the "Anxiety Disorder" section, p. 436.

Substance-Related Disorders [4]. A very long section with a large number of cross references.

You already know I'm not going to talk about Mental Disorders Due to a General Medical Condition. Because the people with these disorders are not crazy.

I'm not going to talk about Substance-Related Disorders either. Even though people with *these* disorders *are* crazy. I assume that substance-related disorders are secondary, a cover—a self-administered "cure"—for other problems.

DSM-IV does not seem to support this view, certainly not in a formal sense. The page reference at Cocaine-Induced Anxiety Disorder, for example, takes you to Substance-Induced Anxiety Disorder (439). The diagnostic criteria distinguish quite clearly a substance-induced from a primary Anxiety Disorder but also make it fairly easy for the clinician to diagnose a primary disorder if he's so inclined.

What's more important from my perspective is that the Diagnostic Features for Substance-Related Disorders offer little information about underlying causes. I'm going to assume that if I know the underlying cause of Anxiety Disorders then I also know the underlying cause of Substance-Induced Anxiety Disorders.

And justified or not I'm going to leap over these three categories to land at Schizophrenia and Other Psychotic Disorders [5]. But only after a detour through fields fragrant with delicate nodding neurotransmitters.

THE PERFORMANCE/BEHAVIOR GAUGE

In the last two to three decades of the twentieth century a number of drugs were found to be effective in the treatment of mental disorders. You know some of the names. You've talked about, perhaps joked about the more popular of these magical medications—Valium, Ritalin, Prozac. Although originally intended for the treatment of a particular type of disorder, most have been used in the treatment of others. Hardly surprising. Not if you believe, dear Reader, as I do, that all mental disorders are much alike in their origins.

Prozac and other antidepressants have been used to treat not only depression but obsessive-compulsive disorder, panic attacks, migraines, anxiety disorders, attention-deficit disorder, kleptomania, post-traumatic stress and even premenstrual mood swings.

The list above and the quotation below are from *Time Magazine*, the issue of May 17, 1999.

> *The reason is that Paxil, Prozac, Luvox and the others all target the same brain chemical, called serotonin, which seems to govern mood. Too little serotonin, and patients tend to feel negative about themselves and the world around them in one way or another. How that dissatisfaction manifests itself—clinical depression, anxiety, phobias, obsessions, even eating disorders—depends on a complex web of factors that researchers have yet to unravel. But they do know that drugs that keep*

*serotonin from being reabsorbed too quickly into the nerve cells—the so-called
selective serotonin reuptake inhibitors, or SSRIs—tend to alleviate these symptoms
(74).*

My imaginary performance and behavior gauge is a function, not a gland or a
ganglia or anything else you could lay your hands on. I'm not going to explain
the nuts and bolts of my measuring device. Or even the myelin sheaths. But you
can bet your life it has something to do with serotonin.

I had thought I would develop my concept of the behavior gauge incremen-
tally, as I made my way from one mental disorder to the next, but now, with
Schizophrenia and Depression just around the corner, I think I should take a
stand, define a position.

I have some notes based on short items found in the Minneapolis *Star Tribune*, *Time Magazine* and Bookshelf 2000. I won't cite specific sources for these
dribs and drabs when it appears they are common knowledge among relevant
professionals. I have another set of notes written some months ago on *Mysteries of
the Mind*, a special issue of *Scientific American*. No date on the cover but published in 1997. I *will* cite two or three papers from this collection.

On the Web I found a massive work entitled *Mental Health: A Report of the
Surgeon General*. I scanned a few paragraphs. You can buy all 2,528 pages at
Amazon.com for 51 bucks. I also discovered a very useful paper by Charles B.
Nemeroff, "The Neurobiology of Depression" in *Scientific American* for June of
1998. I read it online and cannot provide page numbers. But come to think of it
who needs page numbers when you have the FIND ON THIS PAGE command?

In all that follows explanations of obscure phenomena such as the hypothalamic-pituitary-adrenal (HPA) axis are borrowed. Anything that looks like a syn-
thesis is mine.

A few of you know far more about the nervous system than I will ever know,
but most of you know less. To understand mental disorders and their treatment
you have to know what happens at the synapse. So pay attention. But don't make
the assignment more difficult than it really is. The goal is to understand the synapse in the same fuzzy way you understand black holes.

The brain has more nerve cells than you can shake a stick at. And every one of
them can *send* an electrical impulse along its axon, can *receive* impulses through
any of its many dendrites. Details and pictures at *Mental Health*, Chapter 2,
tonight or any night at any time.

Electrons in orbit don't bother me but electrons bumping along a copper wire
make me nervous. I took a course once. Physics. In high school. I can reset a cir-

cuit breaker. Easy. But watts and amperes and potential this and that have never crossed the threshold of my understanding. The next three paragraphs paraphrase material maintained on the Web by the College of Pharmacy, the University of Texas. We have to assume they know what they're talking about.

A resting nerve cell or neuron has a negative charge. There are more negative ions—molecules with an electric charge—inside than outside the axon. The fluid outside the axon has a positive charge and the axon is said to be polarized.

When a neuron is excited or fires sodium ions with a positive charge enter the axon and depolarize it, that is to say they change the electrical charge inside the axon from negative to positive. This change starts at one end of the axon and continues all the way to the other. In response to this electrical impulse (called an action potential) the vesicles—small sacs—swarm to the very edge of the axon and release neurotransmitters—soon to be described—into the synapse—ditto.

After the neurotransmitters are released potassium ions with a positive charge flow out of the axon. Their absence restores the negative charge inside the axon and the neuron is again polarized, at rest, waiting to fire another round.

Don't worry about that action potential, that electrical impulse. Keep your eye on the synapse or synaptic cleft, that exceedingly small space between the axon of one cell, the dendrite of another. On the near shore the electrical impulse becomes a chemical process until, on the far side, the electrical impulse is regenerated and sent on its way, the entire process repeated—on average a thousand times—until the information carried by the impulse—as modified by events at the synapse—reaches its destination, which may be precisely defined by the system. But sometimes not. Much remains unknown.

There are at least a hundred types of neurotransmitter, most or all of which belong to one or another of two groups. In one group—apparently without a general designation—the neurotransmitters consist of relatively small molecules with names now becoming familiar to the general public—dopamine, serotonin and norepinephrine (nor-ep-a-NEF-rin). These three are major components of my behavior gauge.

Neurotransmitters in a second group are based on larger molecules—essentially protein chains—called peptides. I will mention only one of these, corticotropin releasing factor or hormone (CRF or CRH) which is associated with the stress response and the hormone cortisol. CRF and cortisol are *not* components of my behavior gauge but I think I can show a relationship. A logical relationship. No experiments in *my* backyard,

A neurotransmitter cannot roll up anywhere at all on the farther shore like an LST but must bind to a particular protein called a receptor. Almost all known neurotransmitters have more than one, dopamine has five, serotonin at least 14.

Most receptors fall into one of two classes. In one type—electricity again—"a pore within the receptor molecule itself is opened and positive or negative charges enter the cell." If a positive charge the neurotransmitter is excitatory, increasing the likelihood the receiving neuron will fire. If negative, inhibitory, diminishing the chance the neuron will fire. Don't worry about these plebes.

We are concerned with—we *deeply care* about *modulatory* neurotransmitters. We have devoted ourselves to dopamine and serotonin and nor-ep-a-NEF-rin.

These are neither precisely excitatory nor inhibitory but rather act to produce complex biochemical changes in the receiving cell.

The precise brain circuits that carry specific information about the world and that are involved in precise point-to-point communication within the brain use excitatory or inhibitory neurotransmission. Examples of such circuits, which are massively parallel, can be found in the visual and auditory cortex. Overlying this pattern of precise, rapid…neurotransmission are the modulatory systems in the brain that use norepinephrine, serotonin, and dopamine. In each case, the neurotransmitter in question is made by a very small number of nerve cells clustered in a limited number of areas in the brain…

While each of these neurotransmitters is made by a small number of neurons with clustered cell bodies, each sends its axons branching throughout the brain, so that in each case a very small number of neurons, which largely appear to fire in unison when excited, influence almost the entire brain. This is not the picture of systems that are communicating precise bits of information about the world but rather are intrinsic modulatory systems that act via other…receptors to alter the overall responsiveness of the brain. These neurotransmitters are responsible [among other things] *for putting emotional color or significance on top of cold cognitive information provided by* [the precise circuits described above]. *It is no wonder that these modulatory neurotransmitters and their receptors are critical targets of medications used to treat mental disorders…and also are the targets of drugs of abuse.*

And finally, as the impulse moves down the line, some portion of the neurotransmitter left at the synapse is dissipated, the rest returned in a process called reuptake to the sending axon to be reused.

You've seen that word at least once before, in the phrase "selective serotonin reuptake inhibitor," Prozac for one, SSRI for short. A diminished quantity of serotonin is associated with depression. Prozac provides relief by blocking reuptake, thus increasing, for a time, the quantity of serotonin at the synapse.

Ecstasy, the recreational drug of the hour, acts like Prozac run amuck. See *Time* for June 5, 2000, for a cover story that includes a description of neurotransmission.

> *Ecstasy causes the nerve cells to release the serotonin all at once, even without an electrical signal. The chemical floods the synapse, overwhelming the serotonin receptors. Ecstasy also keeps the serotonin from being reabsorbed...*

Which feels great, they say, unless you wake up with a hangover that those who should know call Terrible Tuesday.

I will have more to say about serotonin in this section as I develop the concept of a behavior gauge and again when I comment on depression and the anxiety disorders. I won't have much to say about dopamine until I get to schizophrenia. I hope to establish a link between serotonin and cortisol so I better tell you where it comes from. I don't really care what it is.

The most interesting paper in *Mysteries of the Mind (Scientific American)* is "The Mind-Body Interaction in Disease" by Esther M. Sternberg and Philip W. Gold. They and others in the field have identified an "...intricate network that exists between the immune system and the brain, a network that allows the two systems to signal each other continuously and rapidly." They also find that the immune and stress response systems overlap at many points while together they serve as "the body's principal means for maintaining an internal steady state called homeostasis."

I'm convinced that Sternberg and Gold are on the right track. I wish I could follow them more closely but that's not possible. My excuse now and always is my advanced age and my dedication to squash and old movies. I'm going to assume that whatever Sternberg and Gold say about the immune system will be compatible with my own fanciful conjectures concerning a behavior gauge.

What I need from their paper is a paragraph or two on cortisol.

> *A key hormone shared by the central nervous and immune systems is corticotropin-releasing hormone (CRH); produced in the hypothalamus and several other brain regions, it unites the stress and immune responses. The hypothalamus releases CRH into a specialized bloodstream circuit that conveys the hormone to the pituitary gland, which is just beneath the brain. CRH causes the pituitary to release adrenocorticotropin hormone (ACTH) into the bloodstream, which in turn stimulates the adrenal glands to produce cortisol, the best-known hormone of the stress response (10).*

Or, as we learn from the text under a graphic on the next page:

> *ACTH causes the adrenal glands to release cortisol, the classic stress hormone that arouses the body to meet a challenging situation.*

I trust that you sensed as you read these two passages the nearness of the hypo-thalamus-pituitary-adrenal or HPA axis. But what I really want you to remember is that cortisol is produced by the stress response system and can be measured. You might also file the next quotation where you can easily retrieve it.

> *One of the most widely found biological abnormalities in patients with melancho-lia* [depression] *is that of sustained hypersecretion of cortisol (13).*

> *The excessive secretion of cortisol in melancholic patients is the result predomi-nantly of hypersecretion of CRH, caused by a defect in or above the hypothalamus. Thus, the clinical and biochemical manifestations of melancholia reflect a general-ized stress response that has escaped the usual counterregulation, remaining, as it were, stuck in the "on" position (13).*

Within the next few pages I'm going to challenge the explanation in the para-graph above of the hypersecretion of cortisol and CRH. But I like that switch stuck in the "on" position.

Somewhere toward the beginning of my analysis I said that depression and anxiety go together like Bessie Smith and the blues. So I was very pleased when I found that the Surgeon General concurred.

> *Anxiety and depression frequently coexist, so much so that patients with combina-tions of anxiety and depression are the rule rather than the exception...And many of the medications used to treat either one are often used to treat the other. Why are anxiety and depression so interrelated?*

The Surgeon General's report on *Mental Health* offers no definitive answer but reviews and summarizes the most recent research. I will go directly to Nemer-off's paper on "The Neurobiology of Depression" which covers much the same ground in its first few pages.

The monoamines—you'll recognize some of them, serotonin certainly, dopamine probably, maybe norepinephrine—the monoamines were implicated from the start. Abnormally low levels were clearly associated with depression.

Nemeroff describes the hypothalamic-pituitary-adrenal axis in a passage very similar to that already quoted from Sternberg and Gold, then adds:

> *Chronic activation of the HPA axis, however, may lay the ground for illness and, it appears, for depression.*

Deliberately laboring the point I'm going to present three more quotations bearing on cortisol and related products in connection with depression. You'll see why soon.

> *As long ago as the late 1960s and early 1970s, several research groups reported increased activity in the HPA axis in unmedicated depressed patients, as evinced by raised levels of cortisol in urine, blood and cerebrospinal fluid, as well as by other measures. Hundreds, perhaps even thousands, of subsequent studies have confirmed that substantial numbers of depressed patients—particularly those most severely affected—display HPA-axis hyperactivity. Indeed, the finding is surely the most replicated one in all of biological psychiatry.*

> *Notably, study after study has shown CRF concentrations in cerebrospinal fluid to be elevated in depressed patients, compared with control subjects or individuals with other psychiatric disorders.*

> *Further, postmortem brain tissue studies have revealed a marked exaggeration both in the number of CRF-producing neurons in the hypothalamus and in the expression of the CRF gene (resulting in elevated CRF synthesis) in depressed patients as compared with controls.*

In this context "control" means two groups of people differing in one important respect—diagnosed with depression or not—but matched as closely as possible in all other respects. And that, of course, is as it should be.

But I also see in this situation a wonderful opportunity to bring the methods of biological psychiatry to bear on historical and sociological questions. Almost everyone who has ever written on modernization has had to come to grips with alienation. Yes the word is imprecise, variously defined. But the core meaning is a deprivation of some kind, a loss. And a loss of almost *any* kind can lead to depression, not always to a *diagnosis* of depression, but depression all the same. And maybe elevated levels of cortisol in the urine or the blood or the cerebrospinal fluid.

I do not suggest that we disinter, say, textile workers buried in Manchester in the 1760s in the hope of finding a smidgen of cortisol. But I would like to establish average levels for various groups—Mormons, professors, the sexually confused, textile workers, mail carriers, the Amish, retired folks, blacks and whites, the Intelligentsia, and of course the usual suspects—those earning from $30,000 to $39,999 per year, anyone less than five feet six inches tall, Catholics and Baptists and you get the idea. Good psychiatry. Terrific sociology. Potential dynamite.

Meanwhile Nemeroff has been sketching a theory on the etiology of depression.

> *Neurobiologists do not yet know exactly how the genetic, monoamine and hormonal findings piece together, if indeed they always do. The discoveries nonetheless suggest a partial scenario for how people who endure traumatic childhoods become depressed later in life. I call this hypothesis the stress-diathesis model of mood disorders, in recognition of the interaction between experience (stress) and inborn predisposition (diathesis).*

Note that preemptive "yet." I like a man with confidence. But did it seem to you that Freud's hand must have guided Nemeroff's fingers over the keyboard to produce the provocative phrase "traumatic childhoods?" You'll be surprised to learn I don't want them. Because they might preclude an *initial* onset of depression later in life. Nemeroff knows this.

> *The stress-diathesis model does not account for all cases of depression; not everyone who is depressed has been neglected or abused in childhood.*

But let him finish describing his theory.

> *My colleagues and I propose that early abuse or neglect not only activates the stress response but induces persistently increased activity in CRF-containing neurons, which are known to be stress responsive and to be overactive in depressed people. If the hyperactivity in the neurons of children persisted through adulthood, these supersensitive cells would react vigorously even to mild stressors. This effect in people already innately predisposed to depression could then produce both the neuroendocrine and behavioral responses characteristic of the disorder.*

Nemeroff tested his hypothesis in a series of experiments in which newborn rats were removed from their mothers for brief periods. And sure enough these "maternally deprived rats" as adults showed an increase in the level of corticosterone or rat's cortisol.

In an exciting preliminary finding, Plotsky has observed that treatment with one of the selective serotonin reuptake inhibitors (Paxil) returns CRF levels to normal...

We do not know exactly how inhibition of serotonin reuptake would lead to normalization of the HPA axis.

I don't know either. Not exactly. But in general it looks like the serotonin tells the behavior gauge that all is well, there's nothing to fear. So the stress response is not triggered and the production of CRF and cortisol remains normal.

Nemeroff puts the cart before the horse. He invokes the stress response when he doesn't have a stressor. A traumatic childhood is plausible at first glance but in the end not convincing. Because a good hypothesis has to account for depression in the *absence* of a traumatic childhood.

I would start with a loss. Cumulative maybe. But not necessarily. Maybe suffered in childhood. But mother may have died last Tuesday. Not just any loss will do. There has to be a threshold, but I have no opinion as to what that threshold might be.

The loss produces some or all the symptoms of depression. The symptoms are warnings issued by the behavior gauge telling you that something is wrong, not necessarily with you or your behavior, more likely with the social or even the material world.

The diminished quantity of serotonin is a chemical event within a larger phenomenon that includes the symptoms of depression. Just how the symptoms relate to the serotonin at a chemical or neurological level I have no idea.

If the loss is not made good, the grief assuaged, the symptoms persist. And then the depression *itself* becomes a threat, a danger.

At *this* point and not before the stress response is triggered and the hormones begin to flow as the body is prepared for action, for fight or flight as all the papers remind us. But you cannot fight, you cannot run away from the debilitating effects of chronic depression. The stress response, stuck in the "on" position for as long as the depression lasts, continues to generate anxiety that serves no useful purpose. While cortisol accumulates in the urine and the blood and the cerebrospinal fluid.

And this just in! *Time* has published (01/20/03) a special issue on Mind and Body. Look what I found in an article on depression by Michael Lemonick.

The liver normally produces CRP [C-reactive protein] in response to an immune-system alarm when the body is infected or injured, and CRP is associated with the inflammation that results. For reasons still unknown, though, a recent study of depressed individuals found elevated levels of CRP.

I couda been a contenda. If they wouda let me in the ring.

The behavior gauge functions most effectively in a simple social setting, a primitive setting. While the gauge may have suggested to the intelligent, the impatient, the ambitious—at any time in the long history of the species—a change in *society* rather than a modification in *behavior*, that choice was never the norm.

And now Social Change rattles down the road like a car full of teenagers on Prom Night while Evolution takes her own sweet time. The best behavior gauge available is an old model that still serves some of us fairly well, but for others—too many others—a reuptake inhibitor may be the best bet. If they can live with the side effects, the least of which is a weakened sex drive.

I don't like the drugs. I would rather change the world. I have plans. I know what to do after the last round of grapeshot has been fired. But I don't have the troops. So rest easy, dear Reader. Go in peace.

Loss Continued

And now please turn to the lesson on Schizophrenia. I want you to keep a dictionary within reach as you read. But I know some of you won't. Here's a word I want to be sure you understand.

Prodrome. An early symptom indicating the onset of an attack or a disease. Prodromal. Adjective.

> *The median age at onset for the first psychotic episode of Schizophrenia is in the early to mid-20s for men and in the late 20s for women. The onset may be abrupt or insidious, but the majority of individuals display some type of prodromal phase manifested by the slow and gradual development of a variety of signs and symptoms (e.g., social withdrawal, loss of interest in school or work, deterioration in hygiene and grooming, unusual behavior, outbursts of anger) (282).*

My guess is that most people diagnosed with Schizophrenia struggle as children to meet the standards of the adult world but gain admission more or less on schedule only to find the going too rough, the game not worth the candle. And they quit.

Overwhelmed, in part, by invisible enemies—a genetic predisposition, structural abnormalities in the brain, an abundance of lesser problems (280, 283). Exactly how these factors promote Schizophrenia no one knows, but note, in a list of factors associated with a relatively favorable prognosis, the "absence of structural brain abnormalities" (283).

In the section under discussion, Schizophrenia and Other Psychotic Disorders, there are no other disorders. There are four or five subtypes plus half a dozen ostensible disorders that differ from Schizophrenia proper in one respect only, in duration for example. I'm going to look at the diagnostic criteria for Schizophrenia proper, A1 through A5, then follow them wherever they lead. Two or more are required for the diagnosis.

A1: Delusions. "…erroneous beliefs that usually involve a misinterpretation of perceptions or experiences." Delusions may be more or less plausible—the police have me under surveillance—or bizarre—my thoughts have been taken away by some outside force (275).

At the entry for Schizophrenia five types of delusion are mentioned: persecutory, referential, somatic, religious and grandiose (275).

At the entry for Delusional Disorder, a mild form of Schizophrenia that requires only nonbizarre delusions for the diagnosis, there's another list, with some overlap: erotomanic, grandiose, jealous, persecutory and somatic (296-298).

> *Persecutory delusions are most common; the person believes he or she is being tormented, followed, tricked, spied on, or subjected to ridicule (275).*

> [or perhaps]…*being conspired against, cheated, spied on, followed, poisoned or drugged, maliciously maligned, harassed, or obstructed in the pursuit of long-term goals (298).*

I see the persecutors as parents and teachers, the representatives of the social world. The subtext of the delusion is always the same. They—the people in control—ask too much. They won't let me come inside. They don't love me.

But why a delusion? Why *imagine* persecutors when real enemies are everywhere? Because after the age of 16 or 17—if not much earlier—parents and teachers are not dominant. The authority that imposes restraints is everywhere and nowhere, felt but not always seen. A delusion is a product of the assimilative mode of thought that forges a link between the grievances of childhood and current input. Accuracy is not a requirement.

> *Referential delusions are also common; the person believes that certain gestures, comments, passages from books, newspapers, song lyrics, or other environmental cues are specifically directed at him or her (275).*

Referential delusions are simply a variation on the theme of persecution.

The central theme of delusions based on jealousy is that a "…spouse or lover is unfaithful. This belief is arrived at without due cause and is based on incorrect inferences supported by small bits of 'evidence'" (297).

Betrayed. Again. I'm not loved.

A delusion of the erotomanic type denies rejection. Someone *does* love me.

> *The delusion often concerns idealized romantic love and spiritual union rather than sexual attraction. The person about whom this conviction is held is usually of*

higher status (e.g., a famous person or a superior at work), but can be a complete stranger (297).

God forbid anyone should see in a person of higher status, in a superior, a parent or teacher. And a curse on those who see a wish fulfillment in a tender, romantic attachment to a complete stranger.

A delusion of the grandiose type posits a special talent—I am worthy of love—or a special relationship with a prominent person—I am respected, admired (loved).

And if the prominent person is an imposter, then *I* must be *him*, and you can bet I'm respected, admired (loved) (297).

DSM-IV does not offer an example of a religious delusion but I suspect that grandiose and religious delusions are much alike. If Christ is an imposter, *I* must be *Him*.

Most common [of the somatic delusions] *are the person's conviction that he or she emits a foul odor from the skin, mouth, rectum, or vagina; that there is an infestation of insects on or in the skin, that there is an internal parasite…(298).*

No one loves me. No one could…No one ever will.

Delusions along with auditory hallucinations are characteristic of an important subtype of Schizophrenia, the Paranoid Type. With a few modifications the description could fit any ideologist and a good many political activists. The trick is simply to imbue your delusions with a certain verisimilitude.

The essential feature of the Paranoid Type of Schizophrenia is the presence of prominent delusions or auditory hallucinations in the context of a relative preservation of cognitive functioning and affect.

Delusions are typically persecutory or grandiose, or both…

The delusions may be multiple, but are usually organized around a coherent theme. Hallucinations are also typically related to the content of the delusional theme. Associated features include anxiety, anger, aloofness, and argumentativeness. The individual may have a superior and patronizing manner and either a stilted, formal quality or extreme intensity in interpersonal interactions.

The paragraph that precedes these three quotations now gives me pause. I'll let it stand with the qualification that the diagnosis of choice for your typical ideologist is Bipolar I Disorder, not Schizophrenia. Nevertheless delusions, paranoid delusions "organized around a coherent theme" is a good short definition of an ideology.

(An ideologist is someone like Lenin who plays for keeps, not for votes. Confusion arises because of the casual use of the words "liberal" and "liberalism." There are liberal or free or open practices and institutions going back millennia that have nothing to do with ideology. Parliament is a product of complex historical forces. An ideology is an explanation of the social world—not merely false but delusional—a product of one or several minds under enormous pressure from historical forces. There are many varieties of liberalism. Those that qualify as ideologies should be called by some other name—liberal nationalism under Louis Napoleon, left liberalism or postmodernism in the USA today. Now no more interruptions.)

A2: Hallucinations. Auditory the most common. Usually experienced as voices. Not one's own. "The content may be quite variable, although pejorative or threatening voices are especially common" (275). Just like persecutory delusions. A little crazier.

A3: Disorganized speech taken as indicative of disorganized thought. Speech may follow, for example, a string of "loose associations" and is sometimes, though rarely, so disorganized as to be nearly incomprehensible (276).

A4: Grossly disorganized behavior.

> *Problems may be noted in any form of goal-directed behavior, leading to difficulties in performing activities of daily living such as organizing meals or maintaining hygiene. The person may appear markedly disheveled, may dress in an unusual manner, (e.g., wearing multiple overcoats, scarves, and gloves on a hot day), or may display clearly inappropriate sexual behavior (e.g., public masturbation), or unpredictable or untriggered agitation (e.g., shouting or swearing) (276).*

Another paragraph, still under Criterion A4, describes catatonic motor behaviors:

> *...sometimes reaching an extreme degree of complete unawareness (catatonic stupor), maintaining a rigid posture and resisting efforts to be moved (catatonic rigidity), active resistance to instructions or attempts to be moved (catatonic negativism), the assumption of inappropriate or bizarre postures (catatonic posturing), or purposeless and unstimulated excessive motor activity (catatonic excitement) (276).*

Criteria A1 through A4 are considered positive and are said "...to reflect an excess or distortion of normal functions, whereas the negative symptoms [A5] appear to reflect a diminution or loss of normal functions" (274). There are three of these—affective flattening, alogia and avolition.

> *Affective flattening is especially common and is characterized by the person's face appearing immobile and unresponsive, with poor eye contact and reduced body language (276).*

Reminds me—by way of Poe—of the sullen sad uncertain murmur of a surly student. And also recalls those diagnosed with Conduct Disorder. "They may be callous and lack appropriate feelings of guilt and remorse" (87).

> *Alogia (poverty of speech) is manifested by brief, laconic, empty replies. The individual with alogia appears to have a diminution of thoughts that is reflected in decreased fluency and productivity of speech (276-277).*

DSM-IV doesn't define avolition but a lack of will reminds me of those difficulties related to goal-directed behavior mentioned above.

I see in all this two patterns that are by no means mutually exclusive. In disorganized speech or thought (A3), grossly disorganized behavior (A4), and avolition (A5), especially in connection with goal-directed behavior, I suspect a poorly developed behavior gauge, in part a result of some genetic defect. Not specified. But I'll drop some hints as we move along.

I also see a very stubborn resistance to the constraints imposed by civilization in catatonic motor behaviors, especially catatonic rigidity and negativism (A4), and in affective flattening and alogia (A5). Here too I suspect a poorly developed or defective behavior gauge. A catatonic stupor may even the score with mom and dad but at a price few would wish to pay.

The onset of Schizophrenia typically occurs after the adolescent years but the primary cause of the disorder, I think, is a *virtual* loss, a complete rejection that tends to destroy the minds of its victims, most dramatically in the extreme case of catatonic stupor.

A catatonic stupor approximates death, and in fact some 10% of all those diagnosed with Schizophrenia commit suicide (280).

In a full-blown case of Schizophrenia criteria A1 through A5 might all be satisfied but in most variations on the schizophrenic theme the criteria appear in different combinations. Try these.

The essential features of the Disorganized Type of Schizophrenia are disorganized speech [A3], disorganized behavior [A4], and flat or inappropriate affect [A5]. The disorganized speech may be accompanied by silliness and laughter that are not closely related to the content of the speech. The behavioral disorganization (i.e., lack of goal orientation) may lead to severe disruption in the ability to perform activities of daily living (e.g., showering, dressing, or preparing meals) (287-288).

The silliness and laughter seem childish, regressive, appropriate to the disorganized behavior if not to the content of the speech. Almost as if the behavior gauge were indicating that all is going well when all is going very badly. I think I won't make this assertion now, in connection with Schizophrenia, but I will later, in connection with manic thought and speech.

Schizophrenia, Catatonic Type, is insanity in its most pathetic and repellent form, ringing all the changes on catatonic motor behaviors (A4). I would stress, as I did above, the *resistance* of those who suffer from this disorder to all who try to guide or to help or to *control* them.

There may be extreme negativism that is manifested by the maintenance of a rigid posture against attempts to be moved or resistance to all instructions (288).

Schizophreniform Disorder is Schizophrenia that lasts less than six months. Brief Psychotic Disorder lasts less than one month. There seem to be no disorders that last less than two, three, four, or five months.

Schizoaffective Disorder is Schizophrenia with a Major Depressive, Manic, or Mixed Episode. And finally there is Shared Psychotic Disorder described earlier, the disorder of choice among the Intelligentsia.

Nobody doesn't like Schizophrenia. Sure it's bizarre, repellent, frightening, but also fascinating. Like Hitler and the Nazis.

No one really knows what causes Schizophrenia but recent research has focused on the neurotransmitter dopamine. I didn't learn nearly as much as I had hoped from the University of Texas and the Surgeon General so I went back to the Web where I found *Drugs, Brains and Behavior*, a very useful book by C. Robin Timmons and Leonard W. Hamilton. What I need right now is Chapter 7, "Schizophrenia as a Model of Dopamine Dysfunction." But if you have an idle hour read Chapter 1. Poison arrows. Twitching frog legs. A really good description of the synapse including an explanation of the electrical impulse that I almost understood. The action potential *propagates* down the axon. No circuit. And no batteries.

I fear I've been inconsistent with respect to Web sources. Policy decision right now, January 23, 2001. I won't provide a URL. Because even the best Websites reorganize and change the URL every few hours. I will provide enough information so you can easily find the source. If it's still there.

Test. Use Google. Enter Timmons Hamilton dopamine. Try "I'm feeling lucky" which returns only one item. Bingo. *Drugs, Brains and Behavior.*

Timmons and Hamilton have much of interest to say but they don't know the answer either. It seems likely, they believe, "that some disorder of dopamine metabolism is involved" in Schizophrenia. "…the research findings have shown a continuous, if not entirely coherent, thread of evidence for impaired function of the dopamine systems."

It's curious and I think revealing that Parkinson's disease and Schizophrenia are almost certainly mutually exclusive. Parkinson's disease is associated with a deficiency of dopamine or—in more recent studies—dopamine receptors. Schizophrenia is associated with an excess of one or the other.

Parkinson's disease—sometimes called the shaking palsy—is characterized by muscular tremors and diminished control over movement. Some forms of Schizophrenia, especially the catatonic subtype, are characterized by peculiarities of movement varying from excessive and purposeless motor activity—catatonic excitement—to no movement at all—catatonic stupor. These behaviors are not a result of a loss of control as in Parkinson's disease. While the movements and postures are not voluntary, they are not random either, not simply products of a nervous system gone awry. They carry a message. Frustration. Resistance. Anger. Hate.

Some of you will surely recall that my behavior gauge rests uneasily atop a performance gauge using all the same bodily parts. And the performance gauge for more than a few millennia had to do mostly with bodily movement. It looks to me as if a genetic flaw that involves dopamine metabolism may interfere with, or even disable, the performance and behavior gauge while simultaneously providing a means whereby emotions associated with a virtual loss or losses can be expressed in the form of bizarre and grotesque movements and postures.

> *The initial focus* [of a theory that need not concern us] *was on the behavioral aspects of schizophrenia. The lack of organized and directed behavior, the inability to experience pleasure, and the withdrawal from the external environment all pointed to a dysfunction of the reward system.*

Ignore the reward system. I don't think we need it. Concentrate on the lack of organized and directed behavior. Remember, review what DSM-IV said about goal-directed behavior, about disorganization.

Now the performance and behavior gauge is nothing if not goal-directed. While there is no absolute scale of values the gauge can usually tell you if you're on the right track.

But if the gauge is disabled, presto schizo.

In the section on Mood Disorders {6} as in the section on Schizophrenia there is, I think, only one disorder, but maybe two, if the usual nexus between depression and mania is not preordained. Following DSM-IV I will comment first on *Episodes* which are the defining features of the mood disorders. Although the manual also allows dilute episodes, near- or quasi-episodes that define almost-disorders.

> *The essential feature of a Major Depressive Episode is a period of at least 2 weeks during which there is either depressed mood or the loss of interest or pleasure in nearly all activities (320).*

Along with or as a part of a depressed mood "Many individuals report or exhibit increased irritability (e.g., persistent anger, a tendency to respond to events with angry outbursts or blaming others, or an exaggerated sense of frustration over minor matters)" (321). This is what I would expect if a depressed mood, as I suggested earlier, if depression *itself* is experienced as a threat or danger and triggers the stress response.

A diagnosis requires at least four more symptoms from a list of seven (A3-A9). A sense of worthlessness or guilt (A7) may be of delusional proportions, a reminder of the delusions characteristic of Schizophrenia (A1). Many of those diagnosed with a Major Depressive Episode "report impaired ability to think, concentrate, or make decisions" (A6), difficulties comparable to the disorganized speech (A3)—read thought—and avolition or lack of will (A5) of Schizophrenics.

The psychomotor agitation and retardation (A5) sometimes seen in depression remind me of the disorganized behavior (A4) characteristic of Schizophrenia. Impairment in a depressed state can be almost as crippling as a catatonic stupor.

> *In extreme cases, the person may be unable to perform minimal self-care (e.g., feeding or clothing self) or to maintain minimal personal hygiene (322).*

As you would expect:

> *Major Depressive Episodes often follow psychosocial stressors (e.g., the death of a loved one, marital separation, divorce) (323).*

DSM-IV thoughtfully puts "thoughts of death" last in their list of symptoms (A9). And death in the form of suicide often is the last act for those cursed with either Depression or Schizophrenia.

I don't mean to give too much weight to the similarities between a Major Depressive Episode and Schizophrenia but I *do* have the feeling that the more severe the symptoms are in an instance of *any* disorder the more likely they are to converge with the most severe symptoms of *other* disorders. The ultimate convergence is death by one's own hand.

I just learned at Medscape on the Web (03/19/01) that "Postmortem studies strongly suggest that cerebral concentrations of serotonin are reduced in suicide..."

And this just in! The Minneapolis *Star Tribune,* in an editorial dated April 7, 2001, has proclaimed an anathema on all those mad enough to believe in madness.

> [Suicide is] *the fatal complication of a treatable brain disorder. Nearly all people who die by suicide, neurobiologists note, suffered from diagnosable diseases characterized by low levels of the neurotransmitter serotonin. There's nothing fuzzy about the link: Research shows that 95 percent of suicide victims' brains are serotonin-deficient. And people with low serotonin, the scientists say, are 6 to 10 times more likely to die by suicide...*

> [Suicide is] *the worst symptom of a cluster of disorders typically regarded as mental illnesses. But depression, schizophrenia and bipolar disorder—the three conditions most closely linked with suicide—are every bit as physical as diabetes or cancer.*

Or a broken leg. Because psychological explanations cannot be tolerated. Because God knows where they might lead.

But I still have to ask *why* those diagnosed with depression or schizophrenia or bipolar disorder have lower levels of serotonin than the rest of us. Too much red meat? Not enough exercise?

Maybe a low level of serotonin is the sign of a loss that you and I can scarcely understand. A loss that begs for a change—perhaps an impossible change—in the circumstances of one's life.

Prozac is better than suicide. Better than a shot with a beer chaser. But I don't envy anyone hooked on a reuptake inhibitor.

> *A Manic Episode is defined by a distinct period* [at least one week, less if hospitalization is required] *during which there is an abnormally and persistently elevated, expansive, or irritable mood.*

> *Although elevated mood is considered the prototypical symptom, the predominant mood disturbance may be irritability, particularly when the person's wishes are thwarted. Lability of mood (e.g., the alternation between euphoria and irritability) is frequently seen (328).*

The Associated Features suggest that irritability tends to mean anger and hostility. The *person...DSM-IV never* says the manic-depressive or the schizophrenic. God forbid we should hurt some lunatic's feelings. We must *always* say the *person.* Or the *individual.*

> *The person may be hostile and physically threatening to others. Some individuals, especially those with psychotic features, may become physically assaultive or suicidal (330).*

In addition to an elevated or irritable mood a diagnosis requires three symptoms—sometimes four—from a list of seven (B1-B7). Five of these have to do with the way the person, the individual thinks or expresses himself.

> *Inflated self-esteem is typically present, ranging from uncritical self-confidence to marked grandiosity, and may reach delusional proportions (Criterion B1).*

> *Manic speech is typically pressured, loud, rapid, and difficult to interrupt (Criterion B3).*

> *The individual's thoughts may race, often at a rate faster than can be articulated (Criterion B4)...Frequently there is flight of ideas evidenced by a nearly continuous flow of accelerated speech, with abrupt changes from one topic to another.*

> *Distractibility (Criterion B5) is evidenced by an inability to screen out irrelevant external stimuli...*

> *The increase in goal-directed activity often involves excessive planning of, and excessive participation in, multiple activities (e.g., sexual, occupational, political, religious) (Criterion B6).*

Criterion B7 looks like a corollary of the others:

> *Expansiveness, unwarranted optimism, grandiosity, and poor judgment often lead to an imprudent involvement in pleasurable activities such as buying sprees, reckless driving, foolish business investments, and sexual behavior unusual for the person...*

In my journey through the pages of DSM-IV I've dedicated at least one sentence to almost every mental disorder that has come my way, but I think now I'll abandon that practice. There are two Mood Disorders that really count, Major Depressive Disorder—you need a Major Depressive Episode for the diagnosis—and Bipolar I Disorder—that's right, a Manic Episode for the diagnosis. By the way that's "I" for One, not Insane. As for the remaining six or eight disorders huddled nearby, they remind me of nothing so much as the WPA hard at work.

I've often mentioned—possibly *too* often—that mental disorders regularly overlap both within and across the categories of DSM-IV. Just one more time.

> *The differential diagnosis between **Psychotic Disorders** (e.g., Schizo-affective Disorder, Schizophrenia, and Delusional Disorder) and Bipolar I Disorder may be difficult...because these disorders may share a number of presenting symptoms (e.g., grandiose and persecutory delusions, irritability, agitation, and catatonic symptoms)...(354-355).*

But psychotic symptoms *do* occur in the absence of prominent mood symptoms. So I will shelve the Psychotic Disorders, then pose a fundamental question on the nature of Bipolar I Disorder. What binds Mania and Depression together in an alternating rhythm that has been observed for centuries?

DSM-IV notes the pattern:

> *Roughly 60%-70% of Manic Episodes occur immediately before or after a Major Depressive Episode (353).*

But DSM-IV offers no explanation. Not a hint. Not a clue.

So once again I repaired to the Web where I discovered that the psychiatrists of America had assembled for an exchange of views in May of 2000. A session devoted to "The Spectrum of Bipolarity" drew a standing-room only crowd, while another session—apparently less well attended—addressed the topic of "Complex Bipolarity: Focus on Temperament and Mixed States."

No one offered an explanation of the cozy relationship between mania and depression but the concept of polarity lost some of its luster because recent studies reveal that manic and depressive symptoms often occur simultaneously. While psychotic symptoms lurk in every nook and fissure.

I cannot rest easy without an explanation of the relationship between mania and depression. But why—if the American Psychiatric Association is reluctant to speculate why should *I* court ridicule by spinning a theory from little more than old carpet scraps and a few personal observations? Because *any* hypothesis is better than *no* hypothesis. It gives me something to shoot at. You too, if you're listening.

Let's start with the sense of worthlessness or guilt often associated with a Major Depressive Episode, a sense or feeling that may reach delusional proportions. This symptom stands in sharp contrast to—is the polar opposite of—the inflated self-esteem typically present in a Manic Episode, self-esteem that ranges from "uncritical self-confidence to marked grandiosity, and may reach delusional proportions."

The obvious inference is that a manic episode is an attempt to banish a depressed mood. But the depression is real, the result at least in part of events or conditions in the external world. The behavior gauge is still operating—not disabled as I think it is in the more severe cases of Schizophrenia—but with a distinct bias that has to be characterized as a malfunction. Because *everything* the manic thinks or says or does is given a high positive rating regardless of the impact on the social world, regardless of the possible consequences.

Both the old name and the new—manic-depression and Bipolar I Disorder—fit the description I've just given with its emphasis on polarity. But the names and the description are off the mark. They obscure what the manic-depressive really feels.

He is not a happy man. His elevated mood may be described as "euphoric," even "cheerful," but those who know him, DSM-IV adds, recognize the mood to be excessive. Because at odds with the depression that may be partially eclipsed but is never far away.

> *Mood* [during a Manic Episode] *may shift rapidly to anger or depression. Depressive symptoms may last moments, hours, or, more rarely, days. Not uncommonly, the depressive symptoms and manic symptoms occur simultaneously (330).*

On the assumption that manic behavior is an attempt to overwhelm or escape from depression let's take another look at my model. A loss processed by the performance/behavior gauge produces depressive symptoms indicating that all is not well. If the symptoms persist they come to be interpreted by the system as a threat or danger in themselves. The threat generates anxiety and triggers the stress response.

If the stress response is sustained, if the switch remains in the "on" position—and it will so long as the depressive symptoms persist—then the system, the organism—maybe you, maybe me—finds itself stalled and sputtering in a situation for which evolution has provided no ready solution.

Neither fight nor flight is a realistic option. The enemy's invisible. There's nowhere to go. What options then *are* there for someone in deep depression, someone who's suffered an unbearable loss? Suicide always beckons from afar. Psychiatric help and spiritual guidance are usually at hand but often resisted, evaded.

In the middle range the available options include, in no particular order, a continuation of depressive symptoms for an indefinite period, perhaps in the form of a Major Depressive Disorder. "Approximately 50%-60% of individuals with Major Depressive Disorder, Single Episode, can be expected to have a second episode" (341). And the prognosis gets worse with the second. And the third. And the fourth.

It appears that anxiety may sometimes overwhelm depression, pushing the sufferer toward an Anxiety Disorder.

> During a Major Depressive Episode, some individuals have Panic Attacks that occur in a pattern that meets criteria for Panic Disorder (323).

And finally, in desperation, the switch may be thrown the other way—don't ask me how—so that the behavior gauge somehow, with much grinding and clashing of gears, gives a high positive rating to every word and action to produce a Manic Episode, perhaps a Bipolar I Disorder.

Somewhere along the way I said that the assimilative mode of thought is the only mode in dreams, the predominant mode in mental disorders. In normal waking thought the assimilative mode is checked, more or less kept under control by the empirical mode, but without the assimilative mode, I think, there would be no imaginative thought in either science or the arts.

I mention this because many artists really are crazy. Or so Kay Redfield Jamison claims—with good reason it seems to me—in her article "Manic-Depressive Illness and Creativity" in *Mysteries of the Mind (Scientific American)*. Not everyone who is both unusually accomplished and intense, moody or eccentric is a lunatic Ms. Jamison reminds us. But—

> *All the same, recent studies indicate that a high number of established artists—far more than could be expected by chance—meet the diagnostic criteria for manic-depression or major depression given in the fourth edition of* [you know what] *(45).*

> *These investigations* [described in the article] *show that artists experience up to 18 times the rate of suicide seen in the general population, eight to 10 times the rate of depression and 10 to 20 times the rate of manic-depression and its milder form, cyclothymia (47).*

I have great confidence in Ms. Jamison because she seems to know what I mean by the assimilative mode of thought.

> *Studying the speech of hypomanic patients has revealed that they tend to rhyme and use other sound associations, such as alliteration, far more often than do unaffected individuals. They also use idiosyncratic words nearly three times as often as do control subjects. Moreover, in specific drills, they can list synonyms or form other word associations much more rapidly than is considered normal (48).*

I live for alliteration. Two lines I'm especially proud of—from a satire, *Alice In Washington*—are "oft repeated automotive motifs" and "a coterie of cauldron cognoscenti," lines that actually make sense in context. Proving that *I'm* not crazy. Not unless this shapeless bag of a book is one enormous preternatural manic aberration. But of course it's not, dear Reader. I would tell you if it were.

To qualify as an artist the manic-depressive must find some way to channel the rush of ideas, to give them an aesthetic form. Oddly enough, to qualify as an *ideologist*, the manic-depressive need only shroud the rush of ideas in a pink and golden mist while promising a plenitude of beer and bratwurst to the fevered faithful in some dazzling if phantom future.

What some artists, all ideologists and manic-depressives have in common is an inflated sense of self-importance (Criterion B1). The modern concept of a self-appointed elite distinguished solely by its identification with the *Zeitgeist*, by its foreknowledge of History, rests squarely on this sense of self-importance. Marx in *The Communist Manifesto* put it this way:

> *Finally, in times when the class struggle nears the decisive hour, the process of dissolution going on within the ruling class, in fact within the whole range of society, assumes such a violent, glaring character, that a small section of the ruling class*

cuts itself adrift, and joins the revolutionary class, the class that holds the future in its hands. Just as, therefore, at an earlier period, a section of the nobility went over to the bourgeoisie, so now a portion of the bourgeoisie goes over to the proletariat, and in particular, a portion of the bourgeois ideologists, who have raised themselves to the level of comprehending theoretically the historical movement as a whole.

The Communists [still a portion of the bourgeois ideologists], *therefore, are on the one hand, practically, the most advanced and resolute section of the working-class parties of every country* [that's parties, not the working class], *that section which pushes forward all others; on the other hand, theoretically, they have over the great mass of the proletariat the advantage of clearly understanding the line of march, the conditions, and the ultimate general results of the proletarian movement.*

I mention the ideological elites that appeared in the course of the nineteenth century in the context of Mood Disorders because Thomas Warton in *The Pleasures Of Melancholy* (1747) was way ahead of Marx.

Few know that elegance of soul refined,
Whose soft sensation feels a quicker joy
From Melancholy's scenes, than the dull pride
Of tasteless splendour and magnificence
Can e'er afford.

I once thought I would write a paper or even a book on *The Pleasures Of Melancholy*. I still think I might but I know I never will. I have a three-ring notebook in which I had started to gather data, mostly on allusions and other literary references in the poem. I also have detailed notes for an informal lecture I gave in my seminar on the origin of ideologies at Dalhousie.

What's bothering me right now is that the lines I had in mind but hadn't read in years, the lines quoted a moment ago, are not the blockbuster I was looking for, certainly not for the reader who's seldom wondered about the pleasures of melancholy, never given a passing thought to ideological elites.

Read the poem. It's structurally weak but still an effective evocation of the melancholy mood with hints here and there of a looming manic episode. And if the imagery, the style, the conventions of 18th century verse at first seem placid, then read the poem again. The style is more than adequate for the expression of strong emotion—an outburst of anger, a surge of lust, an urge to kill.

I'm tempted to incorporate the entire poem in my text with an analysis but I don't really have that much to say. Just two more brief passages in an effort to show that the modern concept of a political or ideological elite springs from melancholy, from alienation, from the sense of loss experienced by almost everyone in the Western world since the 18th century but felt most intensely by the Intelligentsia.

> *Ye youths of Albion's beauty-blooming isle,*
> *Whose brows have worn the wreath of luckless love,*
> *Is there a pleasure like the pensive mood,*
> *Whose magic wont to soothe your softened souls?*

The youths in question are the melancholy elect, so defined by the phrase "luckless love." The lines quoted are the first in a section in which the poet acknowledges that the pleasures of melancholy are "imagined joys," a "phantom feast," yet the passage ends on a strangely positive note.

> *These are delights unknown to minds profane,*
> *And which alone the pensive soul can taste.*

Anxiety Disorders {7} are for the hoi polloi.

HAS SOCIAL ANXIETY PUT
YOUR LIFE ON HOLD?
You are not alone.
Social anxiety disorder
affects over 10 million Americans.

From a two page ad for PAXIL in *Time Magazine* (09/27/99). The next quotation—from *USA Weekend* (09/29-10/01/00)—you can take with a grain of salt. Or 500 mg of PAXIL.

> *Anxiety disorders are the No. 1 mental health problem in the United States, affecting at least 19 million people ages 18 to 54 each year, or 13% of adults, according to the National Institute of Mental Health...Anxiety is also the leading mental health problem among the trick-or-treat set, affecting 13 million youngsters ages 9 to 17.*

Panic Attacks and—with a certain qualification to be revealed shortly—Agoraphobia are to Anxiety Disorders as Depressive and Manic Episodes are to Mood Disorders.

The essential feature of a Panic Attack is a discrete period of intense fear or discomfort that is accompanied by at least 4 of 13 somatic or cognitive symptoms. The attack has a sudden onset and builds to a peak rapidly (usually in 10 minutes or less) and is often accompanied by a sense of imminent danger or impending doom and an urge to escape. The 13 somatic or cognitive symptoms are palpitations, sweating [and the like] *(394).*

No one knows the cause of Panic Attacks which often strike without warning but which sometimes appear on cue, provoked or triggered in certain circumstances or situations. In the terminology of DSM-IV an attack, if unprovoked, is *unexpected.* If invariably triggered in certain circumstances the attack is *situationally bound.* If usually triggered but not always, sometimes with a delay, the attack is *situationally predisposed.* DSM-IV also recognizes a *limited-symptom* attack for those efforts that fail to attain the required score of four.

The essential feature of Agoraphobia [sometimes a disorder, sometimes not] *is anxiety about being in places or situations from which escape might be difficult (or embarrassing) or in which help may not be available in the event of having a Panic Attack...(396).*

Psychiatrists who take an interest in the Anxiety Disorders argue among themselves as to whether panic—as in Panic Attack—and general anxiety—as in Agoraphobia—are the same thing.

I learned of this disagreement in "Theories of Panic Disorder," Section II of *Empirically Supported Psychological Treatment of Panic Disorder and Agoraphobia,* Volume 1 of *Psychiatry & Mental Health Clinical Management.* Sorry about the title. I'll call it Sanderson/Regno after the authors, William C. Sanderson and Simon A. Regno. I found both gentlemen as well as their book on the Web at Medscape.

The neurobiologists think that panic and general anxiety are not the same. And there is substantial if not conclusive evidence to support their view. DSM-IV leans cautiously in the direction of the neurobiologists.

The anxiety that is characteristic of a Panic Attack can be differentiated from generalized anxiety by its intermittent, almost paroxysmal nature and its typically greater severity (394).

I see anxiety on a continuum, a touch of nervousness at one end, panic at the other. I don't think it matters much whether the two terms identify phenomena that are exactly the same or slightly different. What *does* matter, and what DSM-IV *does* make plain, is that panic and general anxiety often coexist, often disturb the same mind.

The diagnosis of Panic Disorder requires at least two unexpected Panic Attacks. The diagnosis is always made either With or Without Agoraphobia, which is to say with or without general anxiety. But even if Without Agoraphobia one or more of the Panic Attacks must have been followed by persistent concern (anxiety?) about having additional attacks, and/or worry (anxiety?) about the implications of the attack or its consequences, and/or a significant change in behavior related to the attacks (402).

> *In addition to worry about Panic Attacks and their implications, many individuals with Panic Disorder also report constant or intermittent feelings of anxiety that are not focused on any specific situation or event (398).*

> *Although Agoraphobia may develop at any point, its onset is usually within the first year of occurrence of recurrent Panic Attacks (399).*

Generalized Anxiety Disorder frequently co-occurs with Mood Disorders and with other Anxiety Disorders including, of course, Panic Disorder (433).

Sanderson/Regno review several theories that make a stab at explaining the cause or the source of panic attacks and/or general anxiety. Perhaps the most interesting of these links panic attacks to hyperventilation because the symptoms of one are much like the symptoms of the other. The argument is plausible but even if correct, as critics note, the question remains open whether hyperventilation is primary or secondary to other factors such as fear. You can bet your life I'm going to go with fear. Both panic attacks and general anxiety reflect fear of a threatened loss.

And now let's look at some carefully selected data from DSM-IV to see if it—or they—justify my preemptive strike.

I'm going to place the Anxiety Disorders in one or another of three groups for reasons I hope will become clear as I move along. I include in the first group Panic Disorder With and Without Agoraphobia, Agoraphobia Without History of Panic Disorder, Specific Phobia, Social Phobia—the latest fad advertised as Social Anxiety Disorder—and Generalized Anxiety Disorder. In the second group Posttraumatic Stress and Acute Stress Disorders. In the third, standing alone, Obsessive-Compulsive Disorder.

What do those afflicted with a disorder in the first group fear? Many seem to fear dying, one of the 13 somatic or cognitive symptoms characteristic of Panic Attacks. The manner in which they expect death to strike is of particular interest.

> *Some* [diagnosed with Panic Disorder] *fear that the attacks indicate the presence of an undiagnosed, life-threatening illness...Despite repeated medical testing and reassurance, they may remain frightened and unconvinced that they do not have a life-threatening illness.*

> *...individuals with Panic Disorder often anticipate a catastrophic outcome from a mild physical symptom or medication side effect (e.g., thinking that a headache indicates a brain tumor...(398).*

As far as I can tell from DSM-IV people with Anxiety Disorders do *not* die or commit suicide with anything like the enthusiasm of schizophrenics and manic-depressives. For this reason I would merge the fear of life-threatening illness with the fear of being left alone and helpless, fear of neglect, fear of rejection.

> *...an individual* [with Agoraphobia Without History of Panic Disorder] *may be afraid to leave home because of a fear of becoming dizzy, fainting, and then being left helpless on the ground (403).*

Three or four times while working in my garden at my summer home I was startled by the hiss of a hognose snake. The snakes are not large and not venomous but they look like puff adders. They make me nervous. Otherwise I'm not easily frightened, not for more than a few seconds. Heights, riding a bus, cats, flying, elevators, walking across the agora hold no terrors for me.

And when all is said and done the agora does not frighten those with an Anxiety Disorder either. The danger they seek to avoid, the harm they fear comes from other people. The agora is a marketplace, a meeting place and in *that* sense may pose a threat. Because a meeting is an opportunity to pass judgement. With a word, a look, a contemptuous smile. The anxiety-ridden always expect that smile, that unfavorable judgement, that lack of respect, that threatened loss of love.

> *In feared social or performance situations, individuals with Social Phobia experience concerns about embarrassment and are afraid that others will judge them to be anxious, weak, "crazy," or stupid (412).*

> *Common associated features of Social Phobia include hypersensitivity to criticism, negative evaluation, or rejection; difficulty being assertive; and low self-esteem and feelings of inferiority. Individuals with Social Phobia also often fear indirect evaluation by others, such as taking a test (413).*

> *Children with Generalized Anxiety Disorder tend to worry excessively about their competence or the quality of their performance (433).*

And adults sometimes worry about matters that look like problems left over from early childhood, problems that may well have irritated their parents once upon a time, may well have suggested to tender egos a possible loss of affection.

> The *"panic-like symptoms"* [of Agoraphobia Without History of Panic Disorder] *include…other symptoms that may be incapacitating or embarrassing (e.g., loss of bladder control) (403).*

The shrewd reader will have noticed in the preceding paragraphs the unannounced appearance of the shadowy and insubstantial Unconscious. I can sympathize with the reluctance of the dog and pigeon people to acknowledge the existence of unconscious thoughts—they *are* hard to record—but come on. Some poor devil is quaking with fear as he forces himself to enter an elevator. He can't tell us why he's afraid. He has heard and does not dispute the testimony of a Mr. Otis who claims the elevator in question has completed 103,876 trips without mishap. But still the poor man quakes with fear. Because the elevator is associated in his mind with thoughts or ideas that have been *repressed*, that remain *unconscious*. Unless Mr. Otis was wrong. Or lying.

Sanderson/Regno, who have not yet read my intellectual autobiography, are skeptical about unconscious thoughts.

> Since many of these theories [or psychodynamic formulations] *propose unconscious etiologies for PD, the proposed hypotheses may be difficult to test directly and even more difficult to refute since, given a panic attack, an unconscious etiology may be presumed. Therefore, researchers need to develop measures to test these concepts without presuming their existence solely on the presence or absence of panic symptoms.*

Fair enough. But neither the brain nor the mind is a black box. And right now unconscious thought processes look like the best hypothesis.

Despite their misgivings Sanderson/Regno offer three takes on Freud and Anxiety. I'll pass along the third and latest of these, parts of which—a very few parts of which—I think I can use.

> *Modern psychoanalytic theorists propose that both anxiety and panic attacks are invoked by triggers that are symbolically related to infantile wishes and/or fears. Unconscious or conscious cues become associated with earlier innate psychological and biological threats to the organism, such as castration, separation, or parental disapproval, and serve as a trigger for panic onset in adults. Specifically, when*

defense mechanisms are unable to control the unconscious fantasies linked to these infantile fears, panic attacks develop.

I can usually go along with *infantile wishes* but not in this case, not without the content of the wishes specified.

Fears are crucial here, the fear of rejection, the fear of threatened loss.

I can't accept that much of anything is innate beyond walking readiness and the simple reflexes. I can't even imagine what *innate psychological and biological threats* might be.

When I speak of loss I usually have in mind the loss of respect or affection or love, but the loss of familiar things, the departure from familiar places can have the same significance, the same effect. *Castration*—a rare occurrence in *my* neighborhood—would certainly qualify as a major loss, the threat of *castration* a matter of some concern.

With these qualifications in mind let me try to recast the passage quoted above.

Borrowing what I can from modern psychoanalytic theorists I propose that both anxiety and panic attacks are invoked by triggers that are symbolically related to fears of threatened losses, fears that often but not always originate in early childhood. Such fears, based on events real or fancied, are brought into association one with another through the assimilative mode of thought. Fragments of this loose network may reach consciousness but the pattern, and so the meaning, is repressed, blocked from consciousness. Anxiety associated with this set of ideas may float freely in consciousness, a constant disturbance, but I have to assume that anxiety associated with such a set of ideas may sometimes be repressed, at least in part, until some situation—or cue—because of a prior association, opens a gate or path and releases that anxiety as a situationally bound or predisposed Panic Attack. I also have to assume that repressed anxiety is somehow cumulative, that it may break loose when it reaches a certain level to produce an *unexpected* Panic Attack.

In certain circumstances then, especially with respect to situationally bound Panic Attacks, it would seem that those beset with anxiety have found some relief, have gained some small measure of control that may allow them to get on with their lives. It looks like the *threat* of a loss, which produces anxiety, is not as severe or damaging as an *actual* loss, which produces depression.

I had almost forgot my behavior gauge which must have been monitoring threats of loss all along. And if, as I suggested earlier, loss, followed by prolonged depression, triggers the stress response, it may well be that repeated *threats* of loss, with an assist from the behavior gauge, may *also* trigger the stress response. A Panic Attack may be a stress response gone awry.

And if that's the case there should be an accumulation of cortisol somewhere. DSM-IV is not talking but a certain J. L. Abelson is hot on the trail and the Department of Psychiatry, the University of Michigan, is studying that very question. In two hours on the Web I did not find a quotation that would sweep away all objections, but here's something for the doubters, from *Pharmacological Challenges in Anxiety Disorders* (Lawrence H. Price, Andrew W. Goddard, Linda C. Barr, and Wayne K. Goodman).

> *The pharmacological challenge strategy involves administering a test agent under controlled conditions to elucidate some aspect of biological or behavioral function in the organism being studied.*

That means they drug people to see if they can induce a Panic Attack. And brother can they. Let me count the ways.

> *Targum and Marshall in 1989 (87) administered DL-fenfluramine (60 mg po) to nine drug-free panic patients, nine depressed patients, and nine healthy controls. Panic patients exhibited greater anxiety, prolactin, and cortisol responses than the other groups, with six (67%) experiencing a panic attack.*

> *Tancer and Golden (85) gave fenfluramine (60 mg po) to 21 social phobia patients and 22 controls, observing an elevated cortisol response in the patients.*

I promised a second group of Anxiety Disorders. Here they are.

> *The essential feature of Posttraumatic Stress Disorder is the development of characteristic symptoms following exposure to an extreme traumatic stressor involving direct personal experience of an event that involves actual or threatened death or serious injury...or witnessing an event that involves death, injury, or a threat to the physical integrity of another person...(424).*

The symptoms are what you would expect, bad dreams and a strong desire to avoid anything that might recall the traumatic event. Half those diagnosed with Posttraumatic Stress Disorder completely recover within three months. These people are not crazy.

But what about those whose symptoms persist for twelve months or longer? The symptoms described under Diagnostic Features are less revealing than those under Associated Descriptive Features and Mental Disorders. The latter are commonly seen in connection with an "interpersonal stressor," a dry-as-dust term with grim connotations—childhood sexual or physical abuse, domestic battering, being taken hostage, incarceration as a prisoner of war or in a concentration camp, torture. These symptoms include:

> ...self-destructive and impulsive behavior; dissociative symptoms; somatic complaints; feelings of ineffectiveness; shame, despair, or hopelessness; feeling permanently damaged; a loss of previously sustained beliefs; hostility; social withdrawal; feeling constantly threatened; impaired relationships with others...(425).

In some cases of this kind it may be that the assimilative mode of thought has brought the traumatic event into association with a myriad of less troubling events, again real or fancied, that occurred before the onset of the disorder, but some traumatic experiences in and of themselves—say childhood sexual abuse or torture—may destroy personality beyond the possibility of recovery.

The symptoms characteristic of Posttraumatic Stress Disorder—unlike the symptoms of the Anxiety Disorders in my first group—suggest not the fear or threat of loss but loss endured. And indeed the traumatic event that precipitates the disorder *has* happened and *does* lie in the past. These people are at risk not only for other Anxiety Disorders but for Major Depressive and Substance-Related Disorders.

Acute Stress Disorder is very like Posttraumatic Stress Disorder but the disturbance "...does not persist beyond 4 weeks after the traumatic event." But what if it does? Change the diagnosis. DSM-IV is *practical*.

Hints and traces of delusions seem to lurk between the lines in the section on Anxiety Disorders even though a diagnosis requires that a subject recognize his fears to be excessive or unreasonable. In the case of Obsessive-Compulsive Disorder this requirement tends to break down. Note the odd way DSM-IV expresses it.

> At some point [emphasis added] *during the course of the disorder, the person has recognized that the obsessions or compulsions are excessive or unreasonable (Criterion B) (417).*

But DSM-IV offers the clinician a way out, a Specifier, With Poor Insight, that may be applied when "...the individual does not recognize that the obsessions or compulsions are excessive or unreasonable" (419).

An obsession is an unwanted thought that comes in the night and repeats and repeats in your ear don't you know little fool you...One of the more common obsessions is a fear of contamination, typically through shaking hands or han-dling money. The corresponding compulsion demands repeated washing of the hands, sometimes until the skin is raw.

I see in this behavior a similarity with other Anxiety Disorders, a desire to avoid other people—their hands at least—but the more compelling similarity is with the somatic delusions of Schizophrenia. I fear that I'll be contaminated. Filthy. Covered with insects. Unworthy of love.

Another common obsession is a nagging doubt about something that *must* be checked and checked and checked again. Did I lock the door? Did I mail the check? There has to come a time—and very soon—when no sane person could doubt the door was locked. So what does the obsessive-compulsive person really doubt? I have to guess he doubts he is respected, admired, loved. (One of the most common *compulsions* is to request or demand assurances.)

The other common obsessions mentioned by DSM-IV—a need to have things in a particular order, aggressive or horrific impulses, sexual imagery—do not fit the pattern I discerned in the other two. And this time I'm not going to speculate more than I already have. Because nothing comes to mind. Because the Differen-tial Diagnosis is baffling. Because I really don't want to return to Tourette's Dis-order and the tics (419) from which I fled in terror some months ago.

I *do* think the delusional quality of many obsessions is the best place to start probing. And DSM-IV offers some support for this view.

> *In some individuals with Obsessive-Compulsive Disorder, reality testing may be lost, and the obsession may reach delusional proportions...In such cases, the pres-ence of psychotic features may be indicated by an additional diagnosis of* **Delu-sional Disorder** *or* **Psychotic Disorder Not Otherwise Specified** *(421).*

Obsessed by delusions I queried the Web to see what others might think about Schizophrenia and Obsessive-Compulsive Disorder.

I found a paper from the *Journal of Psychiatry & Neuroscience* (January 1999), "Obsessive-Compulsive Disorder in Schizophrenia: Epidemiologic and Biologic Overlays" (Philip Tibbo, Lorne Warneke).

> *Objective: To examine the co-existence of obsessive-compulsive disorder (OCD) with schizophrenia in terms of epidemiology and overlapping biologic substrates.*

You don't want to read this. Lots of words neither you nor I ever saw before. Nothing conclusive here either. Except that Schizophrenia and Obsessive-Compulsive Disorder *do* often go together. Lots of people think so. Tibbo/Warneke list 90 papers.

At the meeting of psychiatrists mentioned above Yoshio Hirayasu presented a paper on "Management of Schizophrenia with Comorbid Conditions." And yes, of course, obsessive-compulsive behavior is a comorbid condition.

I'll leave it at that. No conclusions. But delusions that derive from fear of a threatened loss are certainly part of the story.

Hysteria is or once was the name of a mental illness, "a neurosis characterized by the presentation of a physical ailment without an organic cause" according to Bookshelf 2000. The word also carries the secondary, derivative meaning of "excessive or uncontrollable emotion."

The word and its connotations are offensive to feminists everywhere because, dear Reader, as you're about to learn if you don't already know, the root of hysteria is the Greek word for womb or uterus. The affliction was thought to be peculiar to those who have wombs. Maybe you. Not me.

We're talking *ancient* Greek, *classical* Greek, so it seems harsh that Freud should take the rap for the womb theory. But on the other hand he *did* write, with Josef Breuer, the notorious *Studies In Hysteria.* He *did* rush into print with the sad and shameful story of Dora in *Fragment of an Analysis of a Case of Hysteria.* He's lucky he wasn't hanged.

DSM-IV has little to say on the history of hysteria so here's what I've gleaned from several sketchy histories on the Web. Everything that used to be loosely termed hysteria appears in DSM-IV under two heads, Somatoform Disorders {8} having to do with physical sensations, and Dissociative Disorders having to do with memory and identity. Two Somatoform Disorders of particular importance are Somatization Disorder, formerly called hysteria or Briquet's syndrome, and Conversion Disorder, formerly hysterical neurosis, conversion type. So there must have been at least one other type. Don't ask *me.*

> *The essential feature of Somatization Disorder is a pattern of recurring, multiple, clinically significant somatic complaints. A somatic complaint is considered to be clinically significant if it results in medical treatment...or causes significant impairment in social, occupational, or other important areas of functioning (446).*

It's not easy to earn the diagnosis. You have to experience pain related to at least four different sites or functions, at least two gastrointestinal symptoms other than pain—nausea and abdominal bloating are typical—at least one sexual or reproductive symptom other than pain, and finally at least one symptom other than pain that suggests a neurological condition which DSM-IV seems to equate in this context with conversion symptoms, anything from a lump in the throat to amnesia.

Two diagnostic requirements are of special interest for those curious about how the mind works.

> *The multiple somatic complaints cannot be fully explained by any known general medical condition or the direct effects of a substance (446).*

> *Finally, the unexplained symptoms in Somatization Disorder are not intentionally feigned or produced...(446).*

A comment under Associated Laboratory Findings suggests that one of the editors may have a sense of humor.

> *Laboratory test results are remarkable for the absence of findings to support the subjective complaints (447).*

Symptoms galore for no discernible reason, yet the symptoms are not feigned. Looks like unconscious thought processes to me.

But let's look at Conversion Disorder before we jump to any conclusions.

> *The essential feature of Conversion Disorder is the presence of symptoms or deficits affecting voluntary motor or sensory function that suggest a neurological or other general medical condition (452).*

A diagnosis of Conversion Disorder demands conversion symptoms. But so does a diagnosis of Somatization Disorder. And you must *not* diagnose Conversion Disorder if the conversion symptoms occur exclusively during the course of Somatization Disorder (446, 448-449, 452, 456). They lost me somewhere in that tangle of words.

No matter. Just remember that Conversion Disorder has to do with motor or sensory functions—paralysis or blindness for example, *hysterical* paralysis or blindness if you'll pardon the expression—while Somatization Disorder has to do with sensations, with imaginary aches and pains. Otherwise the two disorders are much alike. And both depend on unconscious thought processes. Just ask DSM-IV.

> *Traditionally, the term* conversion *derived from the hypothesis that the individual's somatic symptom represents a symbolic resolution of an unconscious psychological conflict, reducing anxiety and serving to keep the conflict out of awareness ("primary gain"). The individual might also derive "secondary gain" from the conversion symptom—that is, external benefits are obtained or noxious duties or responsibilities are evaded. Although the DSM-IV criteria set for Conversion Disorder does not necessarily imply that the symptoms involve such constructs, it does require that psychological factors be associated with their onset or exacerbation (453).*

Apart from this passage DSM-IV offers no opinion and few clues as to what meaning might plausibly be attributed to the symptoms characteristic of Somatization and Conversion Disorders. So I will turn, as before, not to Freud but to Menninger. Both disorders look like focal suicide, pain inflicted or a function impaired without actual physical injury.

Undifferentiated Somatoform Disorder and Pain Disorder are obviously similar to the two disorders just described and require no further comment. Not so, I think, in the case of Hypochondriasis and Body Dysmorphic Disorder.

> *The essential feature of Hypochondriasis is preoccupation with fears of having, or the idea that one has, a serious disease based on a misinterpretation of one or more bodily signs or symptoms (462).*

Those who suffer from Hypochondriasis cannot be persuaded they're not sick. Negative test results count for nothing. DSM-IV assures us that the beliefs and fears of those afflicted with this disorder are not of delusional intensity but provides a Specifier, With Poor Insight, that suggests to me their beliefs and fears most assuredly *are* of delusional intensity.

In any case they're really crazy, constantly badgering doctors, sometimes deluding themselves to the extent that they become invalids without benefit of illness. And they frequently suffer from other mental disorders, selected usually from the Anxiety and Depressive categories.

I see, of course, the similarities with Somatization Disorder but I also see in an imagined illness something very like a somatic delusion with a twist, a desperate response to a threatened loss. I'm terribly sick. Pay attention. Don't leave me. Love me.

> *The essential feature of Body Dysmorphic Disorder...is a preoccupation with a defect in appearance (Criterion A). The defect is either imagined, or, if a slight physical anomaly is present, the individual's concern is markedly excessive (466).*

In this case DSM-IV does *not* provide the Specifier With Poor Insight but *does* indicate, under Associated Features and Disorders, that ideas of reference—delusions let me remind you—that:

> *Ideas of reference related to the imagined defect are also common. Individuals with this disorder often think that others may be (or are) taking special notice of their supposed flaw, perhaps talking about it or mocking it (466).*

> *Avoidance of usual activities may lead to extreme social isolation. In some cases, individuals may leave their homes only at night, when they cannot be seen, or become housebound, sometimes for years...The distress and dysfunction associated with this disorder, although variable, can lead to repeated hospitalization and to suicidal ideation, suicide attempts, and completed suicide (467).*

While here, as in Hypochondriasis, I can see the similarities with Somatization Disorder it still seems to me that a somatic delusion is at the focus. I'm ugly. Deformed. No one could possibly love me.

Is it true, I wonder, that if you've pondered the meaning of a word or looked up an unfamiliar word in the dictionary you will surely see that word in print within the next 48 hours? Probably not.

But on the morning after I wrote my last paragraph on Body Dysmorphic Disorder here's what I read in the *Star Tribune* for March 11, 2001, page F15, from a review of Joyce Carol Oates, *Faithless: Tales of Transgression.*

> *Alice, the cynical young waitress in 'Ugly,' is mad at people in general and men in particular. She's convinced that she's repulsively unattractive and therefore unlovable.*

I was not surprised to learn that Body Dysmorphic Disorder may be associated with Delusional and Obsessive-Compulsive Disorders along with Social Phobia and Major Depression.

Looking to the future DSM-IV has provided space for any number of Factitious Disorders {9} but fortunately at the present time there's only one. Because I'm growing weary. Insanity is losing its once undeniable fascination.

> *The essential feature of Factitious Disorder is the intentional production of physical or psychological signs or symptoms.*

> *The motivation for the behavior* [and this is a diagnostic requirement] *is to assume the sick role (471).*

Some of these people spend their entire lives trying to gain admittance to a hospital, or to stay there once admitted.

> *Individuals with the chronic form of this disorder may acquire a 'gridiron abdomen' from multiple surgical procedures (473).*

Since the symptoms are feigned Factitious Disorder might seem at a glance to differ significantly from Hypochondriasis—that must be why DSM-IV provides the disorder with its own special niche—but for the life of me I can't see much difference. The permanent patient is begging for attention. Take care of me. Look after me. Love me.

In their introductory remarks on the Dissociative Disorders {10} the editors of DSM-IV suggest, perhaps inadvertently, that there's a *norm* for human behavior from which at least some mental disorders represent a deviation.

> *The essential feature of the Dissociative Disorders is a disruption in the usually integrated functions of consciousness, memory, identity, or perception of the environment (477).*

I wish they had told us more about these "usually integrated functions" but they didn't and I don't expect they ever will. So when we draw near the end of this short section on the Dissociative Disorders I'll make a stab at explaining exactly what is dissociated from what.

> *The essential feature of Dissociative Amnesia is an inability to recall important personal information, usually of a traumatic or stressful nature…(478).*

> *Dissociative Amnesia most commonly presents as a retrospectively reported gap or series of gaps in recall for aspects of the individual's life history (478).*

DSM-IV writes that the *gaps* are usually related to traumatic or stressful events, but of course they meant to say the *content* of the gaps, that is to say *repressed memories* are usually related to events of that nature.

The memory loss in Dissociative Amnesia is reversible. DSM-IV does not explain how the content is recovered but offers examples that belong to one or another of two distinct types.

Events of the less frequently occurring type, an instance of which DSM-IV calls a "florid episode with sudden onset," would include—I'm guessing now—combat experience, rape, torture and the like. Amnesia for events of this kind looks like a symptom of Posttraumatic Stress Disorder, but maybe not. It's their manual.

What interests me is the *other* distinct type of stressful event, a category that includes "episodes of self-mutilation, violent outbursts, or suicide attempts." Following Menninger I would describe acts of this kind as focal suicide tending toward actual suicide. Dissociation, whether of a memory as in Dissociative Amnesia or a part or function of the body as in Somatization and Conversion Disorders, tends to isolate the problem and may prevent something worse from happening.

> *The essential feature of Dissociative Fugue is sudden, unexpected, travel away from home or one's customary place of daily activities, with inability to recall some or all of one's past (481).*

This looks like a more severe form of Dissociative Amnesia with similar underlying causes.

> *Depression, dysphoria, grief, shame, guilt, psychological stress, conflict, and suicidal and aggressive impulses may be present (481).*

Long before I knew about DSM-IV I was vaguely aware that the concept of a Multiple Personality Disorder was highly controversial. It was easy to find evidence of the controversy on the Web. MPD is overdiagnosed many say, a rare condition if it exists at all. But MPD is very common say others, definitely underdiagnosed.

Paul McHugh, whose review article on DSM-IV I cited earlier, has a brief paper on the Web with the title *Multiple Personality Disorder.*

> *MPD is an iatrogenic behavioral syndrome, promoted by suggestion, social consequences, and group loyalties. It rests on ideas about the self that obscure reality, and it responds to standard treatments.*

Iatrogenic means induced in a patient by a physician's activity, manner, or therapy. Or as McHugh puts it more bluntly in his next paragraph, MPD is "created by therapists."

DSM-IV, which renames MPD Dissociative Identity Disorder, indicates an awareness of the controversy but seems to think the diagnosis may be appropriate at least some of the time. I have to go with McHugh but I'm insensitive enough, rude enough to say that anyone easily led to believe he has multiple personalities is really crazy. And since DSM-IV is in the label business I can go along with Dissociative Identity Disorder.

VideoHound gives *The Three Faces Of Eve* (1957) three and a half bones. I watched it on the VCR not too long ago (02/03/00). Joanne Woodward is really good. And who would not willingly place his life in the capable hands of Lee J. Cobb? I should add, in deference to Dr. McHugh, that two fine performances do not a valid diagnosis make. Keep your eye on the blue china cup.

And by the way:

Self-mutilation and suicidal and aggressive behavior may occur (485).

Depersonalization Disorder is the last in this small set and I think I'll approach it indirectly, through a more general look at dissociation. As used in DSM-IV—and in the psychiatric literature in general I should think—dissociation means that some part of the whole is cast adrift to become more or less autonomous. The "usually integrated functions" are no longer integrated. The individual loses control of some part of his being in a peculiar, unintelligible way, not as a result of an injury or other intrusion from without but as if an internal connection had been severed.

When it comes to particulars it's not easy to say what a dissociated part of the whole might be. So let me offer a simplified view of Somatization Disorder. The diagnosis begins with complaints, various pains here and there in the body, physical sensations it would seem, cause unknown, presumed psychological. The clinician is persuaded of their subjective reality. His patient has come to him in good faith, feigns nothing.

His patient also complains of vomiting or diarrhea, hardly feigned but again no cause can be detected.

I've said more than once that unconscious thought processes are involved but that doesn't tell us much. So please allow me an x factor. For a page or two only. I'll describe the attributes of x, then try to identify it, place it in some larger context.

X has the power to generate physical sensations, whether at the appropriate bodily location or only within the central nervous system I don't know. If at the appropriate location then x has the power to present the sensations to consciousness.

X also has the power to block information from consciousness. This is not immediately obvious in the case of Somatization Disorder but is plainly what happens in the case of Dissociative Amnesia.

And finally x has the power to activate or inhibit various processes and functions outside the central nervous system.

Let's look first at the second attribute, the power to block information from consciousness. Because it's easier to talk about than the others. Ordinarily we can easily recall the important events of our lives. It is precisely this ability that is lost in Dissociative Amnesia and Fugue.

So we're talking about a shift of control from wherever you want to locate it—mind, self, psyche, ego—to somewhere else. To the x factor.

Now let's look at the third attribute, the power to inhibit a motor function, say to paralyze an arm. With the right subject a good hypnotist has the same power. And DSM-IV notes that "hypnotizability" is characteristic of all the Dissociative Disorders.

Many years ago—maybe 1949—I watched a fine performance by amateur hypnotist Red Grant, an older brother of my high school classmate Bill Grant, a lawyer by training I think, but then an officer in the navy home on leave. He joined half a dozen young men and their dates at a small party during which he hypnotized a young woman, 17 or 18 years old, an eager subject, not too bright. Red patiently explained in a soft deep baritone that her body was growing stiff and rigid from the top of her head to the bottoms of her feet. And so it was. I helped suspend that body between two chairs, only the heels and the back of the head touching, She was straight and stiff as a two by six across a pair of sawhorses. It's a great trick.

Obviously the hypnotist gains a large measure of control over his subject. There must be something about the nervous system that makes this possible. My guess is the x factor.

In a book I mentioned in passing—*The Girl Of My Dreams*, still unpublished—I developed certain arguments that I have need of right now. I posited a large entity, a "single idea" as the functional unit of the mind. I would define the single idea, if pressed, as the sum of electrical pulses in the central nervous system at any given moment. In an effort to sound even more scientific I would identify the single idea with the macropotential waveform or waveshape.

I also posited a primary psychic location—a set of single ideas—a location that does not actually hold anything—because memories are not stored—but that has the power to reference the more important experiences of early childhood. The location may be marked by a tag or label or "symbol"—a single idea—that may appear in a hypnogogic reverie as you drift off to sleep.

A dream starts from the point at which some fragment from recent sensory input is assimilated to something in the primary psychic location. In the broadest sense a dream is a reconstruction of some part or all of the primary psychic location using new materials gathered from more recent memories accessed by the assimilative mode of thought.

The reconstruction requires two steps. The first is the unconscious dream, the generation during NREM sleep of a string of ideas or images based on a theme or set of related themes from the primary psychic location and linked by superficial similarities.

The second step is the generation during REM sleep from the pool of ideas accumulated during the unconscious dream of the dream proper, the dream as dramatic fragment, the dream sometimes recalled upon awakening. I attributed the creation of this dream to a dreamwright, at first simply as a literary convenience, a name for a process I could not explain at a deeper level. But I felt from the beginning that in some strange way the dreamwright worked independently of the dreamer. I was finally driven to conclude that the dreamwright and the primary psychic location are intimately related, part of the same phenomenon.

And they *do* act independently of the dreamer. There is no homunculus hiding in the folds of the cerebral cortex but rather something more like a ghost, the semblance of a child, stunted, denied direct access to the external world, his only recourse the assimilative mode of thought, his only aim the recreation of himself and the minuscule world of his infancy through a kind of metaphorical invasion of the mind of the person he has become.

The dreamwright and the primary psychic location constitute the x factor with the power to produce not only dreams but symptoms as varied as a lump in the throat or hysterical blindness. It's unfortunate, I concede, that I have no data to support this fascinating hypothesis. But I have to wonder why those diagnosed with Depersonalization Disorder see the world as a dream.

And there you have it. Simple. Neat. Plausible. Maybe not compelling. But better than anything you'll find in DSM-IV.

DSM-IV places the Sexual and Gender Disorders {11} in one of three subcategories—Sexual Dysfunctions, Paraphilias and Gender Identity Disorders.

In anticipation of a run through the Sexual Dysfunctions (493) DSM-IV describes a sexual response cycle in four phases—desire, excitement, orgasm, resolution. The disorders that follow are pegged to the phases but to no valid end so far as I can see.

Whatever the diagnosis the problem is always much the same, an unwillingness shading into an inability to perform or to complete the act of love. Because of a lack of desire (Hypoactive Sexual Desire Disorder), an aversion to the very thought (Sexual Aversion Disorder), an inability to perform (Female Sexual Arousal Disorder, Male Erectile Disorder), incomplete or unsatisfactory performance (Female Orgasmic Disorder, Male Orgasmic Disorder, Premature Ejaculation), and obstacles to performance, sometimes insurmountable (Dispareunia or pain associated with intercourse, Vaginismus or severe contractions when penetration is attempted. Or even contemplated).

Certain crucial events in the sexual response cycle are involuntary, not subject to the will even at the best of times. The Sexual Dysfunctions tend to cluster around these involuntary responses which may be of interest from a medical point of view, but not, I think, from a psychological point of view. Most of the Sexual Dysfunctions look like Conversion Disorder. That Conversion Disorder affects *voluntary* motor or sensory functions does not suggest a different etiology. (I understand that psychiatrists are physicians and must take into consideration a great many things I am free to ignore.)

Meanwhile you might also note that Dispareunia bears a striking resemblance to Pain Disorder. Dissociation is at work.

I used the phrase "act of love" a few paragraphs back rather than "coitus" or some such term because we're talking primarily about love and its loss, about sex derivatively. Those unwilling or unable to perform or to complete the act of love have mixed feelings about any loved person, negative feelings predominating. Some loss in the past has spread itself abroad through the assimilative mode of thought to contaminate all potential love relationships. The Sexual Dysfunctions are variations on the theme of focal suicide.

DSM-IV departs from its usual format with respect to the Paraphilias (522). The term—the plural form—is given the full treatment, Diagnostic Features, Prevalence, Course and so on. But the plural form will not do as a diagnosis.

The actual disorders are briefly described *without* any of the usual subdivisions. The first paragraph opens not with the typical introductory phrase—the essential feature of this or that disorder—but with a special formula—the paraphiliac focus in, say, Exhibitionism...The formula suggests that all the Paraphilias are fundamentally alike. And so I believe, dear Reader, although I'm hard pressed to say exactly why. We'll just inch along for a page or two. Very carefully.

Those troubled by a Sexual Dysfunction usually have in mind a more or less normal sexual relationship. With a human being. Still breathing. Of the opposite sex. Old enough to see over the steering wheel.

By way of contrast those afflicted with a Paraphilia tend to abjure anything that in the least resembles a normal sexual relationship. A stolen pair of panties may be someone's cup of tea. Or a whip across bare buttocks.

DSM-IV describes eight Paraphilias and names seven more under Paraphilia Not Otherwise Specified. The secondary seven occur less frequently than the essential eight and are probably, from the perspective of the Man in the Street, more ghoulish, more repulsive—necrophilia, coprophilia—but not fundamentally different.

The preferred psychoanalytic view is that the Paraphilias "represent a regression to or a fixation at an earlier level of psychosexual development." (I'm quoting from one of dozens of guides and glossaries on the Web.) If this is so—and I think it's part of the explanation—then I infer that the paraphiliac has been rejected at an early age. Nothing in his experience since early childhood has been more rewarding than whatever satisfaction he derives from his peculiar behavior.

There's another psychoanalytic theory according to which the Paraphilias "are all expressions of hostility in which sexual fantasies or unusual sexual acts become a means of obtaining revenge for a childhood trauma." I like this one. But it needs some tweaking. Hold on.

Six of the essential eight Paraphilias involve another person or persons. The other two—Fetishism and Transvestic Fetishism—attribute a special significance to some object, usually an article of clothing, which implies another person.

At least half the Paraphilias involve behavior that may harm other people. That harm may be limited to embarrassment or an emotional shock (Exhibitionism, Frotteurism) but often involves physical injury, even death (Pedophilia, Sexual Sadism).

With the obvious exception of Sexual Masochism the Paraphilias do not seem to lead to self-mutilation or self-punishment.

On the other hand nearly all the Paraphilias seem to *invite* punishment by flaunting the criminal law. I'm tempted to say that paraphiliacs more or less deliberately provoke punishment, but "provoke" is too strong a word. Paraphilliacs have to know they're at risk but they don't seem to care. Newspaper accounts suggest they resume their illegal behavior as they walk out the gate. They rarely seek help from the medical profession or anyone else.

I see the paraphiliac as someone who has been rejected at an early age, who has suffered, in my terminology, a severe virtual loss. He vents his anger on an introjected object which should lead to focal suicide but—except in the case of Sexual Masochism—does not. Somehow the process is reversed. A match for the introjected object is found in the external world. A stranger will do although the person selected may have to meet certain criteria with respect to age and gender. The reversal makes it possible to harm a real person other than the self through an introjected object. I have to assume that the original loss occurred at such an early age that the conscience remained rudimentary at best. In many cases of Pedophilia and Sexual Sadism I would say that the self has triumphed over conscience, has blown by the restraints of civilization, become a monster.

The description of the behavior of Sexual Sadists in DSM-IV and elsewhere justifies, I think, the use of the word *revenge*. Someone has to pay for the rejection.

Sadism and Masochism are opposites in a formal, logical sense, but they often coexist in the same psyche. I have a case history to suggest that at least some sadists are punished in the end.

John Osborne's play *Look Back In Anger* opened at the Royal Court in London not long before my arrival as a student in the fall of 1956. Osborne was the most important of several "angry young men" who generated a fair amount of intellectual excitement for the next few years. Kenneth Tynan skillfully rode the wave generated by Osborne, first to a career as theater critic, then to a position he created for himself, dramaturge for the new National Theater. I used to read his reviews, first in the Sunday paper, *The Observer*, then in *The New Yorker*.

Tynan is no longer with us but he left behind some journals, excerpts from which were published in two issues of *The New Yorker*—August 7 and 14, 2000—with a brief introductory essay by John Lahr. Tynan, the journals reveal, was both a spanker and a spankee.

> *Since last November I have been seeing (and spanking) a fellow spanking addict, a girl called [Nicole, a pseudonym]. Her fantasy—dormant until I met her—is precisely to be bent over with knickers taken down to be spanked, caned or otherwise punished, preferably with the buttocks parted to disclose the anus. She also enjoys exposing and spanking me.*

> *The apprehension, the preparation, the threat, the exposure, the humiliation—these are thrilling, and so is the warmth afterwards, and the sight of the marks; but the impact of pain on bottom is no fun at all.*

> [Pain] *is the unpleasant price that must be paid for the pleasure that precedes and follows it.*

Tynan apparently felt no guilt about his peculiar pastime which made him, he seemed to believe, a much more interesting person than drones like you and me. Sado-Masochism is "infinitely more varied in its excitements" than straight sex.

But he played right into Menninger's hands. Here's a line—quoted by Lahr—that Tynan wrote during psychoanalysis.

A bedwetter, I soiled my mother and she punished me by refusing to feed me.

Tynan smoked. And kept right on smoking against advice from all quarters. He died of pulmonary emphysema at the age of 53. Menninger would say he smoked himself to death. And so would I.

The very old among you may remember a time when the sexually confused included homosexuals, when homosexuality was a mental disorder. No more. In a speech before the assembled psychiatrists of North America in 1973 Sir Isaac Newton, in a stunning reversal, announced that henceforth all scientific disputes would be settled by majority vote after a period of intensive lobbying. A clinician who encounters "Persistent and marked distress about sexual orientation" *is* permitted to diagnose Sexual Disorder Not Otherwise Specified.

The third and last subcategory in this section is Gender Identity Disorders. There's only one. You can guess the name. Boys diagnosed with this disorder generally try to act as if they were girls, girls as if they were boys. And this despite the fact that they all have normal genitalia. Adults with the disorder try to look like someone of the opposite sex. Many try to pass in public.

All of them—boys and girls, men and women—invite trouble which the society at large is more than willing to provide.

Many individuals with Gender Identity Disorder become socially isolated. Isolation and ostracism contribute to low self-esteem and may lead to school aversion or dropping out of school (534).

Children with Gender Identity Disorder may manifest coexisting Separation Anxiety Disorder, Generalized Anxiety Disorder, and symptoms of depression. Adolescents are particularly at risk for depression and suicidal ideation and suicide attempts. In adults, anxiety and depressive symptoms may be present (535).

The comments under Associated Physical Examination indicate practices hard to distinguish from self-mutilation, all too often in my view surgeon assisted self-mutilation. This, and the trouble they bring on themselves, the predilection for suicide suggest to me that Menninger's focal suicide might be the best explanation.

But the ostensible goal of people with this disorder is not to harm or punish themselves. They want to change their sexual identity. Why?

You already know I have no conscience when it comes to spinning theories but I'm ordinarily reluctant to invent data. But hey, in this case DSM-IV is not giving me much. And it's easy to imagine a child overhearing Mommy or Daddy say, "I wish we'd had a boy instead of Suzy. Why wasn't Tommy born a girl?"

DSM-IV *does* offer some hard evidence.

> *By late adolescence or adulthood, about three-quarters of boys who had a childhood history of Gender Identity Disorder report a homosexual or bisexual orientation, but without concurrent Gender Identity Disorder (536).*

Long before I waded into the psychiatric swamp it seemed to me that homosexuals strive to reenact their childhood following a revised script. The best of them are children in adult bodies.

DSM-IV describes two Eating Disorders {12} that are very much alike.

> *The essential features of Anorexia Nervosa are that the individual refuses to maintain a minimally normal body weight, is intensely afraid of gaining weight, and exhibits a significant disturbance in the perception of the shape or size of his or her body (539).*

The anorexic typically loses weight by eating much less than the body requires but DSM-IV provides a subtype—Binge-Eating/Purging—that brings Anorexia much closer to the other Eating Disorder, Bulimia Nervosa.

The bulimic always binges and usually purges although other methods of keeping the weight down are sometimes employed. Whatever the method the bulimic always maintains a weight more or less within the normal range.

Eating problems that do not necessarily meet the criteria for a diagnosis are apparently very common.

> *WESTPORT (Reuters Health) Mar 23—Nearly one third of high school girls and 16% of high school boys show symptoms of an eating disorder, researchers told participants of the annual meeting of the Society for Adolescent Medicine, in San Diego, California, on Thursday* [March 22, 2001].

According to Dr. S. Bryn Austin of Children's Hospital in Boston, reporter for the research team at the conference:

> *Twelve percent of the girls and 4% of the boys surveyed said that they vomited to control their weight; and 7% of the girls and 6% of the boys reported binge-eating at least once per week.*

If any of these teenagers are significantly less attractive than the norm it's because they are emaciated, not because they're too fat. Like those who suffer from Body Dysmorphic Disorder they *persuade themselves* they are ugly. Therefore not worthy of love. Therefore not loved.

But appreciated, I should think, by psychoanalysts everywhere. Orality right out in the open for all to see. Mewling and puking as existential angst. DSM-IV offers no clues but I'll bet my life—well no, but maybe a case of Oppenheimer Krotenbrunnen auslese—that the dreams of anorexics and bulimics are rife with evidence for introjection.

Some of them die at an early age. "Of individuals admitted to university hospitals, the long-term mortality from Anorexia Nervosa is over 10%" (543).

DSM-IV organizes the Sleep Disorders {13} into four sections, three of which I will ignore for reasons given many long months ago—Sleep Disorder Related to Another Mental Disorder, Sleep Disorder Due to a General Medical Condition, and Substance-Induced Sleep Disorder. I'll have to devote some time and space to the Primary Sleep Disorders, but not much. Because I'm getting old. Because I've spent too much time thumbing through DSM-IV. Because Insanity is driving me crazy.

> *Primary Sleep Disorders are presumed to arise from endogenous abnormalities in sleep-wake generating or timing mechanisms, often complicated by conditioning factors. Primary Sleep Disorders are in turn subdivided into **Dyssomnias** (characterized by abnormalities in the amount, quality, or timing of sleep) and **Parasomnias** (characterized by abnormal behavioral or physiological events occurring in association with sleep, specific sleep stages, or sleep-wake transitions (551).*

The phrase "endogenous abnormalities" suggests to me physical rather than mental or psychological problems, and that would seem to be the case with Circadian Rhythm Sleep Disorder, "…a persistent or recurrent pattern of sleep disruption that results from a mismatch between the individual's endogenous circadian sleep-wake system on the one hand, and exogenous demands regarding the timing and duration of sleep on the other" (573).

Several of the Sleep Disorders, however, seem to have a large mental or psychological component. In the case of Primary Insomnia, for example, "Symptoms of anxiety or depression that do not meet criteria for a specific mental disorder may be present" (553-554). "Individuals with Primary Insomnia may have a history of mental disorders, particularly Mood Disorders and Anxiety Disorders" (554).

DSM-IV warns against making the diagnosis if the insomnia is not the predominant complaint, "sufficiently severe to warrant independent clinical attention" (556). I have to wonder if Primary Insomnia can ever be a satisfactory diagnosis. A better choice might be Insomnia Related to Another Mental Disorder.

People with Primary Hypersomnia sleep too much. Some doze off during normal waking hours while others sleep as long as twelve hours a night. DSM-IV implies that the problem is psychological but offers no explanation. And strange to say Etiola, the mischievous tenth daughter of Mnemosyne and Zeus, has failed to provide me with an insight into the excessively drowsy. But please note that "…many individuals with Primary Hypersomnia have symptoms of depression that may meet criteria for Major Depressive Disorder" (559).

The essential features of Narcolepsy are repeated irresistible attacks of refreshing sleep, cataplexy [temporary loss of muscle tone], *and recurrent intrusions of elements of rapid eye movement (REM) sleep* [dream sleep] *into the transition period between sleep and wakefulness (562).*

According to the International Classification of Diseases Narcolepsy is a neurological condition but DSM-IV codes the disorder on Axis I reserved for crazies. Those who suffer from this peculiar illness don't look crazy to me but "A concurrent mental disorder or history of another mental disorder can be found in approximately 40% of individuals with Narcolepsy" (564). The disorders most commonly associated with sudden attacks of sleep are the Mood Disorders, Major Depressive and Dysthymic.

The nightmare in Nightmare Disorder is a bad dream. Bad enough to bring you bolt upright in your bed, heart pounding, palms clammy. Bad enough to cause, in the carefully chosen words of DSM-IV, "significant subjective distress" but usually nothing worse. Since nightmares occur frequently in conjunction with any number of mental disorders (582) I see Nightmare Disorder at the midpoint of a continuum, ordinary dreams at one end, Major Depression and Schizophrenia at the other.

From the corner in which I seem to have positioned myself I have to say that the *content* of nightmares, properly analyzed, would reveal a loss, actual or threatened.

Now as it happens some forty years ago I actually analyzed four or five dreams. In the most transparent of these a robbery is taking place although it's not clear who's stealing what. The language in which the dream is transcribed is a thicket of b's and p's, a discovery of the greatest importance fully explained in *The Girl Of My Dreams*. What concerns us now are the two b's in the middle of robbery, letters that for a time concealed a baby, my little sister, in fact, whose birth robbed me of the attention that was my due. In brief a loss.

But I do not think I need to claim that a loss can be discovered in every dream, much less every slip of the tongue. I suspect—and now I'm making things up again—that a wide range of psychological problems only tenuously linked, if at all, to a loss may figure in dreams, but only a grave loss, actual or threatened, can produce symptoms that meet the criteria for the diagnosis of a severe mental disorder.

Sleep terrors are "abrupt awakenings from sleep" but those who experience sleep terrors, DSM-IV tells us in the same paragraph, are difficult to awaken (583). I know not what lies behind this odd error but my guess is that those who experience sleep terrors *look* like they're awake to those aroused by a scream in the night even though they don't know what's happening. Even though next morning they remember next to nothing.

Sleep terrors, the essential feature of Sleep Terror Disorder, are accompanied by "behavioral manifestations of intense fear" but occur during NREM sleep so there is no dream, no story or imagery that might explain the fear. My first thought was an unexpected nocturnal Panic Attack but DSM-IV notes that Panic Attacks during sleep "produce rapid and complete awakening without the confusion, amnesia, or motor activity typical of Sleep Terror Disorder" (586).

A short passage in the Associated Features seems more revealing than anything under Diagnostic Features.

> *The individual may actively resist being held or touched, or even demonstrate more elaborate motor activity (e.g., swinging, punching, rising from the bed, or fleeing). These behaviors appear to represent attempts at self-protection or flight from a threat…" (584).*

The lines quoted describe a stress response. I suggested above that persistent depressive symptoms might trigger a stress response but I see no sign of that here. Sleep terrors look like a false alarm, the product of faulty wiring and some other event, possibly a stressful emotional event from the recent past but more likely some trivial physiological occurrence.

Sleep terrors are experienced by children far more often than by adults. And they seem little the worse for it.

> *Children with Sleep Terror Disorder do not have a higher incidence of psychopathology or mental disorders than does the general population (584).*

> *Sleep Terror Disorder usually begins in children between ages 4 and 12 years and resolves spontaneously during adolescence (585).*

And note that sleep terrors seem to be the *cause* of the usual "clinically significant distress or impairment." Who wants to sleep with someone who may scream bloody murder in the middle of the night?

Sleepwalking Disorder is like Sleep Terror Disorder without much terror, but note that "running and frantic attempts to escape some apparent threat" *can* occur (588). And so the differential diagnosis *can* be difficult.

The first of the Impulse-Control Disorders Not Elsewhere Classified {14} is Intermittent Explosive Disorder, a reference to temperament, not an inordinate fondness for bombs. I have spent enough time with DSM-IV to know at a glance that "intermittent" means "sort-of-intermittent." I sense the inevitable "may."

> *Signs of generalized impulsivity or aggressiveness may be present between explosive episodes (610).*

Intermittent Explosive Disorder is rare and must be carefully distinguished from a score of others. But why bother? I'm going to call it Sort-Of-Conduct Disorder and refer you to my earlier remarks.

(Susan L. McElroy—*Recognition and Treatment of DSM-IV Intermittent Explosive Disorder*—found that 25 of 27 subjects recruited for her study had been diagnosed with a Mood Disorder. The favorable response of symptoms to treatment with mood stabilizers suggests to Dr. McElroy "the possibility that intermittent explosive disorder may be linked to bipolar disorder." Wouldn't surprise me. But then it wouldn't surprise me if explosive episodes were linked to hot humid days and sleepless nights. But seriously folks, take a look at Associated Physical Examination Findings and General Medical Conditions, at developmental difficulties, at histories of neurological conditions. I still like Sort-Of-Conduct Disorder.)

> *The essential feature of Kleptomania is the recurrent failure to resist impulses to steal items even though the items are not needed for personal use or for their monetary value (612).*

The impulse to steal is experienced as ego-dystonic, a term that appears several times in the text of DSM-IV but not in the index or glossary. The dictionaries immediately available to me, online and ondesk, do not define it but the term is all over the Web, in part, I think, because the label "ego-dystonic homosexuality" slipped past the thought police a few years ago. I inadvertently Asked The Expert at Mental Health Interactive what it means and I liked his answer.

> *Some clinicians view kleptomania as part of the obsessive-compulsive spectrum of disorders, reasoning that many individuals experience the impulse to steal as an alien, unwanted intrusion into their mental state. Other evidence suggests that kleptomania may be related to, or a variant of, mood disorders, such as depression.*

Hold tight to the concept of an ego-dystonic action while we take a quick look at compulsive shopping, not recognized by DSM-IV but similar to kleptomania in the eyes of Dr. Susan McElroy, a prominent figure at the 153rd Annual Meeting of the American Psychiatric Association. She and others have found low levels of serotonin in compulsive shoppers and kleptomaniacs.

And now, Ladies and Gentlemen, watch closely as I shuffle the cards. We have on display people who buy or steal things they have no use for. The impulse to buy or steal feels strange, as if the decision were not their own, as if they were acting—my own small imaginative contribution—in a dream. Nevertheless they take pleasure in the act despite some fear of apprehension on the part of the kleptomaniacs, despite an awareness of wrongdoing, a sense of guilt.

We know our shoppers and stealers have low levels of serotonin. And they *may* have been diagnosed, according to DSM-IV, with Major Depressive Disorder (612). From my perspective they have suffered a loss. And now *you*, dear Reader, see as plainly as do I they're trying to buy or steal whatever it was they lost. They are necessarily disappointed in every purchase, every theft, but they seldom abandon the quest despite, in the case of kleptomaniacs, "multiple convictions for shoplifting" (613).

Somewhere on the Web I read that kleptomaniacs unconsciously want to be caught and punished. Maybe so. But it's the pursuit of the ineffable that keeps them going.

Pathological Gamblers, like kleptomaniacs and compulsive shoppers, are trying to recover, *desperately* trying to recover something lost. DSM-IV notes that pathological gamblers may be "overly concerned with the approval of others" (616). And of course increased rates of Mood and other Disorders have been reported in individuals diagnosed with Pathological Gambling.

I see the same forest fires on the tube as you do, but I don't believe I've ever witnessed a fire raging out of control. We used to burn leaves in the street in Grand Rapids but you can't do that anymore. I've never set fire to a house.

So I was surprised to learn how common it is to start fires that are a threat to property and life. Those most likely to set a fire are children and adolescents who, if caught, are usually diagnosed, if at all, with Conduct Disorder, not Pyromania.

True Pyromania—to set a fire without regard to the consequences and without apparent motive beyond the gratification derived from the fire itself—true Pyromania is rare. And DSM-IV is not about to hint at a possible explanation.

I found a paper on the Web with the simple title *Pyromania* by a John E. Hamling. He edits a journal called *Nereus* and may be a psychoanalyst. His papers look scholarly but I haven't found any letters after his name. I'm going to cite his paper anyway. What the hell, there are those who would question even *my* credentials in psychiatry.

> There have been a number of articles written over the years on pathological fireset-ting and it is obvious from reading these accounts that pathological firesetters are not an homogeneous group. In an earlier paper (Hamling, 1995) I suggested using Freud's psychosexual stages as a means of subgrouping these people...Of all the groups, the two that seem to meet the DSM-IV...criteria for pyromania are oral-stage firesetters and phallic-stage firesetters...

> Oral-Stage Firesetting.

> The attraction of fire to oral-stage firesetters seems to be its subconscious relation-ship to the warmth and security of maternal care. There is often maternal neglect in the first 18 months of these people's lives [six citations follow].

I like this, of course, because it fits so well with my predilection for The Explanatory Loss. I'm not even going to tell you about phallic-stage firesetters.

Trichotillomaniacs pull out their own hair by the roots from every part of their bodies. They do not "routinely" report pain from this remarkably bad habit but man it's gotta hurt. Yet pain or no pain they derive some sort of pleasure or gratification from the actual pulling of hair even though their lives may be a shambles.

The chief similarity in the disorders described in this section of DSM-IV is compulsive behavior, not impulsive behavior despite the name for the set. You might go to the movies on an impulse. Or buy a pizza. You would not systematically remove your eyebrows one hair at a time.

There's an element of self-destruction in all these disorders, most obviously so in Pathological Gambling and Trichotillomania. The latter is plainly self-mutilation and Menninger's terms chronic suicide and focal suicide are both applicable.

I find on the Web there are support groups for those who suffer from Trichotillomania. Which is fine. God knows they need help. Some of them, perhaps most of them spend their lives in a strange self-imposed isolation trying to hide their bizarre behavior from prying eyes. Listen to Christina Pearson quoted by Arline Kaplan in "Trichotillomania: Out of the Closet" (*Psychiatric Times,* November 1997).

> *The absolute suffering and the desire to die rather than expose myself is what drove me to try to make a difference. It was the absolute agony that I experienced as a teenager and through my entire 20s. I was tying my hands together, taping my mouth shut, wearing ski masks to bed—terrified that someone would discover I was defective. I was pulling my hair out, eating it, and not understanding that if I was so smart, why couldn't I stop. I lost all of my higher education to trichotillomania. I dropped out of eighth grade. I closed the door to intimate relationships. I was spending six to eight hours per day pulling my hair and trying to run a small business.*

As much as I sympathize with Christina Pearson and others who suffer as she has I'm tired of reading in the *Star Tribune,* on the Web, almost everywhere I turn that there are no crazy people anywhere. Certainly not in America. Here's what I think. And if I lose the respect of Ms. Landers that's a cross I'll have to bear.

If you tear your hair out by the roots you are crazy. Or insane. Or nuts or bonkers. The word doesn't matter. You are out of your mind.

If a steady diet of Prozac curbs the urge to pull out your hair that's wonderful. God bless you. But you're still crazy. And crazy you shall remain until you can drop the pills and leave the hair alone.

Adjustment Disorder {15} is a "residual category" (625) that differs from the others in that a diagnosis requires "an identifiable psychosocial stressor or stressors" (623). Strangely enough there is no list of symptoms under Diagnostic Features beyond the usual requirement of distress or impairment. To make the diagnosis, however, the clinician must choose one of six available Subtypes and these suggest that Adjustment Disorder approaches but does not meet the criteria for a Mood, Anxiety or Conduct Disorder. By definition Adjustment Disorder cannot last more than six months after the precipitating stressful event but nevertheless is associated—no Subtype specified—with an increased risk of suicide attempts and suicide (624). I have to consider Adjustment Disorder a close relative of Major Depression.

Under Recording Procedures DSM-IV notes that, "In a multiaxial assessment, the nature of the stressor can be indicated by listing it on Axis IV (e.g., Divorce)" (624). There must be a ton of data scattered in offices across the country that a sociologist or historian could make good use of. I assume, by the way, without any data at all, that if the distress or impairment is severe enough to warrant the diagnosis then the stressor has probably been brought into association with other psychological problems by the assimilative mode of thought.

There are ten Personality Disorders {16} organized in three clusters, all ten coded on Axis II, not with the "more florid Axis I disorders" (26). I'm not going to talk about any of them. Because they're all Sort-Of or Not-Quite or Almost Disorders. They don't offer a new point of departure for an understanding of mental illness.

DSM-IV knows it. And actually concedes the point in the Differential Diagnosis (632-632) and in a special subsection on Dimensional Models for Personality Disorders.

> *The diagnostic approach used in this manual represents the categorical perspective that Personality Disorders represent qualitatively distinct clinical syndromes. An alternative to the categorical approach is the dimensional perspective that Personality Disorders represent maladaptive variants of personality traits that merge imperceptibly into normality and into one another (633).*

> *The DSM-IV Personality Clusters (i.e., odd-eccentric, dramatic-emotional, and anxious-fearful) may also be viewed as dimensions representing spectra of personality dysfunction on a continuum with Axis I mental disorders (634).*

That's it. I hope you know more now than you did before. I do. Before I started I knew dreams, I think, and ideologies, but for the rest I was just feeling my way. Until I wrote the section on the performance and behavior gauge. Until

I explained serotonin and the synapse to *myself*. Until I pulled an etiology for Major Depression out of a hat.

I'm not saying I've got it right. I'm telling you that sometimes a long shot pays off. I'm a Pathological Gambler. With ideas. Not money. Long as they are I like the odds on *this* bet.

But win or lose, right or wrong, the high excitement of the past ten months is mine to keep. That's somebody else you hear singing *I got them bruisin' broodin' where you gone to baby I need serotonin blues.*

0-595-28204-0

www.ingramcontent.com/pod-product-compliance
Lightning Source LLC
Chambersburg PA
CBHW031233280526
45784CB00004B/1550